Carbon Dioxide Equilibria and Their Applications

JAMES N. BUTLER
HARVARD UNIVERSITY

Carbon Dioxide Equilibria and Their Applications

ADDISON-WESLEY PUBLISHING COMPANY
Reading, Massachusetts   Menlo Park, California
London   Amsterdam   Don Mills, Ontario
Sydney

Library of Congress Cataloging in Publication Data

Butler, James N.
  Carbon dioxide equilibria and their applications.

  Bibliography: p. 2
  Includes index
  1. Carbon dioxide.   2. Carbonates.   3. Chemical
equilibrium.   I. Title.
QD181.C1B87   546'.6812   81-3494
ISBN 0-201-10100-9   AACR2

ISBN 0-201-10100-9
ABCDEFGHIJ-AL-898765432

Preface

When I wrote *Solubility and pH Calculations* nearly 20 years ago, I had in mind an audience of analytical chemistry students. Long after I finished that book, I discovered that some of my chief fans were geochemists and environmental engineers. From their point of view, the most important acid–base and solubility problems were missing from *Solubility and pH*. These were the equilibria of carbon dioxide and the solubility of carbonates, which in 1963 I considered too complicated for an elementary course. This book is an attempt to remedy that omission and to provide enough detail from this important cluster of equilibrium problems so that a diligent student can master its complexities.

Carbon dioxide occupies a central place in the biosphere and in many of the geological processes that create and erode rocks. Plants, from trees to phytoplankton, absorb CO_2 from the atmosphere and convert it into biomass; respiration by terrestrial and aquatic plants and animals returns carbon dioxide to the atmosphere. Many marine plants and animals convert CO_2 into calcium carbonate, and when they die the mineral portions of their bodies become reefs, sediment, and limestone. Thus a full understanding of the natural environment involves a quantitative understanding of the transformations of carbon dioxide and carbonate minerals. Recent concern has centered on how rapidly the the excess CO_2 produced by humans burning coal and other fossil fuels can be assimilated by biological and geochemical processes.

Almost all natural waters contain carbonate and carbon dioxide. Water conditioning and wastewater treatment must therefore include processes involving acid–base equilibria of CO_2 and precipitation or dissolution of $CaCO_3$. These topics are considered an essential part of the environmental engineering curriculum, but often are not given a very detailed treatment in engineering textbooks.

This book is about the details of carbonate equilibrium calculations and how they are applied to oceanography, geochemistry, and environmental engineering. It begins at a level accessible to most students with a sound quantitative course in general chemistry and a reasonable mastery of algebra at the high-school level. As the book proceeds, more and more complicated examples are considered, but I have tried to give correspondingly more detailed discussion.

My intent was to provide a textbook monograph that would be useful in a variety of courses. These would certainly include the aquatic chemistry courses of the environmental engineering curriculum, courses in geochemistry and chemical oceanography, and quite possibly other courses I have not yet imagined. It is not meant to substitute for a textbook in any of these subjects, but ought to provide enough supplementary material to satisfy the curiosity (and perhaps alleviate the confusion) of all but the most advanced students. Indeed, I hope this book will be useful to students who would find the introductory chapters far too elementary, but who need to understand the details of ionic strength corrections or the quantitative theory behind the lime–soda water softening process.

I have taught much of this material since 1970 in advanced undergraduate and graduate courses at Harvard. Every year, my lectures metamorphosed as I learned new perspectives on what seemed at first to be a straightforward topic. I hope some of that sense of adventure remains in this book.

I would probably not have considered writing this book if it had not been for the encouragement of Tom Robbins at Addison-Wesley, and would not have completed it except for the inspiration given by my colleagues Robert Garrels, H. D. Holland, Fred Mackenzie, James J. Morgan, Raymond Siever, the late Lars Gunnar Sillen, Werner Stumm, and Roland Wollast. Christine Lawton and Cora Bennett typed the manuscript with unusual skill and patience. A lot of my thinking and writing was done at the Bermuda Biological Station, where carbonate geochemistry has a long tradition. My wife Rosamond has given me moral support when no one else could.

Finally, I want to dedicate this book to my friend Conrad D. Gebelein (1944–1978), who taught me much of what I know about the real world of carbonate geochemistry. His untimely death was a great loss to his friend and to science.

Cambridge, Mass.   **J. N. B.**
December, 1981

Contents

Review of Solubility and pH Calculations

This book is intended to be self-contained, but of course it depends on knowledge of some chemistry and mathematics. The topics reviewed in this chapter are presented in more detail, with more examples, in my other books.* My intent here is only to give a catalog of what you ought to be familiar with for the rest of this book.

CONVERTING CHEMICAL MODELS TO MATHEMATICAL PROBLEMS

If you are going to calculate the equilibrium composition of an aqueous solution, you must have in mind a "chemical model" of the system. A chemical model consists of a set of chemical species and the equilibria relating them. This information may not be explicitly stated in a problem, and a successful attack on the problem may require some prior chemical knowledge on your part. Part of the model, of course, is the set of numerical values for the appropriate equilibrium constants. In this book, I have tried to make my chemical models explicit.

Equilibrium constant expressions alone are not enough to provide an answer to most problems, and additional relations between the concentrations of species are needed. I feel that the method of mass and charge balances is easiest to understand and to generalize, even though it may lead to a little more algebra than other methods† in some of the simplest examples. The combination of equilibria with mass and charge balances will give you as many independent equations as you

---

* See Butler, J. N. 1964. *Solubility and pH Calculations* or *Ionic Equilibrium*. Reading, Mass: Addison-Wesley. I refer to these books throughout; for brevity, they will be cited by their titles only.

† You may have learned a method normally limited to those relations that can be derived from one balanced chemical reaction. There is also a method that involves minimization of total free energy, which is the basis of some computer programs but is not feasible for hand calculation.

have unknown species, and you can then solve this set of simultaneous mathematical equations.

LOGARITHMIC AND EXPONENTIAL FUNCTIONS

Because the numerical values of concentrations in aqueous solutions range from 10 M down to less than 10^{-15} M, logarithmic functions of the variables are common and graphical representations are usually made on a logarithmic scale. The function you are probably most familiar with is

$$pH = -\log_{10}[H^+]\gamma_+,$$

where $[H^+]$ is the concentration of hydrogen ion in moles per liter and γ_+ is its activity coefficient, a correction for non-ideality which is usually between 0.7 and 1.2 but depends on the concentrations of other ions in the solution.*

For some examples in this book, the activity coefficient is approximated by $\gamma_+ = 1.0$ for simplicity, and

$$pH = -\log_{10}[H^+].$$

For other examples, I will show you how to make numerical estimates of the activity coefficients.

A NOTE ON CALCULATORS

In the numerical work presented in my examples, concentrations are sometimes written as $z = 5.28 \cdot 10^{-4}$, and sometimes as $10^{\log z}$ or $z = 10^{-3.28}$. (These are numerically equal, as you can verify.) You should have a calculator that has both "$\log_{10}x$" and "10^x" function keys for the most efficient use of these notations. Practice by doing numerical examples like this:

$$z = 10^{-3.28} + 10^{-3.55} - 10^{-4.20} = 10^{-3.13}.$$

Regardless of your calculator's logical system, you should be able to evaluate such expressions without writing down any intermediate answers.

Note that, if only two decimal places are retained in the logarithms, any terms at least two orders of magnitude less than the largest can be neglected without calculation. For example:

$$10^{-3.28} + 10^{-5.28} = 10^{-3.276} \cong 10^{-3.28}.$$

You can see that addition of the second term affected the answer only in the third decimal place of the logarithm.

---

* See Chapter 2; see also *Ionic Equilibrium*, Chapter 12, or Bates, R. G. *Determination of pH: Theory and Practice*, ed. 2. 1973. New York: John Wiley Interscience.

STRONG ACIDS AND BASES

When water ionizes, a proton (hydrogen ion) is transferred from one water molecule to another, resulting in a hydrated hydrogen ion and a hydroxyl ion:

$$H_2O + H_2O \rightleftharpoons H_3O^+ + OH^-.$$

Actually, H_3O^+ is further hydrated by at least three additional water molecules, and the complexity of the structure increases at lower temperatures. Because this hydration structure is not normally part of the chemical models discussed in this book, I write the hydrated proton as H^+ (a common simplification); but you should keep the complexities in the back of your mind.

The equilibrium constant expression for the ionization of water in its most general form would be written

$$[H^+][OH^-]\gamma_+\gamma_- = K^0[H_2O]\gamma_0,$$

but in dilute solutions the activity coefficients are close to 1.0 (see Chapter 2) and the concentration of water is nearly constant at $[H_2O] = 55.5$ mole/L. Therefore, the concentration ion product

$$[H^+][OH^-] = K_w = \frac{K^0[H_2O]\gamma_0}{\gamma_+\gamma_-} \qquad \textbf{(1.1)}$$

is commonly used.* Both the activity of undissociated water and the activity coefficients of the ions can be included in the equilibrium constant K_w. In dilute aqueous solutions at 25°C, $K_w = 10^{-14.0}$. In more concentrated solutions, K_w depends on the concentrations of other ions (see Fig. 2.8 for more details).

The second equation required to solve for pH is a *charge balance* between H^+ and OH^-, which states that all positive ions formed must be balanced by an equal number of negative charges. In pure water,

$$[H^+] = [OH^-]. \qquad \textbf{(1.2)}$$

Substitute the ion product of water (1.1) and you get

$$[H^+]^2 = K_w.$$

Taking the negative logarithm of both sides and setting $K_w = 10^{-14.0}$ and $\gamma_+ =$ 1.0, you get:†

$$\text{pH} = -\log[H^+] = -\frac{1}{2}\log K_w = 7.0.$$

2 log[H+] = log Kw
log [H+] = ½ log Kw
substitute in def. pH = −log[H+]
pH = −½ log Kw = 7.0
neg. log − not needed

* In some books K_w is used for the activity product $a_H a_{OH} = [H^+][OH^-]\gamma_+\gamma_-$. I use K_w for the concentration product at finite ionic strength and K_w^0 for the activity product, since K_w^0 is equal to the concentration product K_w extrapolated to zero ionic strength. At 25°C, $K_w^0 = 10^{-13.999}$.

† Note that if you set $\text{pH} = -\log([H^+]\gamma_+)$ and $\gamma_+ = \gamma_-$, you will find $\text{pH} = \frac{1}{2}\text{p}K_w^0$ even if γ_+ or γ_- is not 1.0.

Addition of acid to pure water will increase $[H^+]$; addition of base will decrease $[H^+]$ and increase $[OH^-]$. If an acid is fully dissociated into ions it is called *strong*; an example is HCl, which yields only H^+ and Cl^- in aqueous solution.

For a solution containing some constant C molar HCl, two equations are required besides the ion product of water—a *mass balance* on chloride (all the chloride in solution comes from HCl):

$$[Cl^-] = C; \tag{1.3}$$

and a *charge balance*, including chloride as well as hydroxyl ion:

$$[H^+] = [OH^-] + [Cl^-]. \tag{1.4}$$

Equations (1.1), (1.3), and (1.4) can be solved without approximation. Substitute from the mass balance (1.3) into the charge balance (1.4) to eliminate $[Cl^-]$, then solve the resulting equation for $[OH^-]$:

$$[OH^-] = [H^+] - C. \tag{1.5}$$

Substitute Eq. (1.5) in the ion product (1.1) to obtain a quadratic equation:

$$[H^+]([H^+] - C) = K_w = 10^{-14.0}. \tag{1.6}$$

Normally, C is large compared to 10^{-7}, and this should encourage you to neglect $[OH^-]$ compared to $[Cl^-]$ in the charge balance (1.4) and to try the approximation*

$$[H^+] = [Cl^-] + \cdots = C + \cdots. \tag{1.7}$$

EXAMPLE   Find pH for 10^{-4} M HCl. With $C = 10^{-4}$, approximation (1.7) gives pH = 4.0. Substitution in the full quadratic (1.6) gives

$$[H^+]([H^+] - 10^{-4.0}) = 10^{-14.0}$$

$$[H^+] = 10^{-4.0} + \frac{10^{-14.0}}{[H^+]} = 10^{-4.0} - 10^{-10} = 10^{-4.0}.$$

Here I have substituted the approximate value in the second term on the right, to show that it was negligible.   ∎

You will notice, of course, that when C is small compared to 10^{-7}, the result is pH = 7.0, the same as in pure water.

---

* The missing term $[OH^-]$ is represented by three dots to remind you that this result is an approximation. In other examples or approximate equations the quantity represented by "$+ \cdots$" may be much more complicated, and may be either positive or negative. In the example, you see that "$+ \cdots$" is -10^{-10}, which is negligible compared to $10^{-4.0}$.

A strong base (such as NaOH) is completely ionized in solution, and the charge balance for a solution containing both strong base and strong acid is

$$[H^+] + [Na^+] = [OH^-] + [Cl^-],\qquad (1.8)$$

and the mass balances are

$$C_a = [Cl^-]\qquad (1.9)$$

$$C_b = [Na^+],\qquad (1.10)$$

where C_a and C_b are the molar concentrations of acid and base respectively. Proceeding as above, you can combine Eqs. (1.8), (1.9), and (1.10) with the ion product (1.1) to give the quadratic

$$[H^+]([H^+] + C_b - C_a) = K_w.\qquad (1.11)$$

The same sort of approximations apply. When there is excess strong acid,

$$[H^+] = C_a - C_b + \cdots.\qquad (1.12)$$

When there is excess strong base ($C_b > C_a \gg [H^+]$), Eq. (1.11) reduces to

$$[H^+] = \frac{K_w}{C_b - C_a} + \cdots,\qquad (1.13)$$

which is what you would derive by setting the hydroxyl ion concentration equal to the excess of strong base:

$$[OH^-] = C_b - C_a + \cdots\qquad (1.14)$$

and substituting in the ion product (1.1).*

WEAK ACIDS

When an acid does not fully dissociate in solution, it is called "weak," and description of such solutions requires the use of the familiar equilibrium between the undissociated (HA) and dissociated forms:

$$HA \rightleftharpoons H^+ + A^-,$$

which gives the following equilibrium expression:

$$[H^+][A^-] = K_a[HA]$$

or

$$10^{-pH}[A^-] = K_a'[HA],\qquad (1.15)$$

---

* The titration of a strong acid with a strong base, or vice versa, is discussed briefly in Chapter 3 of this book, and in *Ionic Equilibrium*, Chapter 4.

where again the activity coefficients have been included in the equilibrium constant.* The mass balance for total weak-acid concentration C is

$$C = [HA] + [A^-] \tag{1.16}$$

and the charge balance for pure water containing only the weak acid is

$$[H^+] = [A^-] + [OH^-]. \tag{1.17}$$

Equations (1.15), (1.16), and (1.17), together with the ion product of water (1.1), comprise four equations in four unknowns, and their general solution is straightforward (see *Solubility and pH Calculations*, pp. 46–47 or *Ionic Equilibrium*, pp. 116–120).

These equations are easily represented by a logarithmic concentration diagram, such as shown in Fig. 1.1. This is a plot of the various concentrations as a function of pH (in the figure, I have set $\gamma_+ = \gamma_- = 1.0$):

(handwritten: 1.18 from definition of pH)

$$[H^+] = \frac{10^{-pH}}{\gamma_+} = 10^{-pH} + \cdots, \tag{1.18}$$

(handwritten: 1.19 is 1.11 with def. of pH substituted for [H⁺])

$$[OH^-] = \frac{K_w 10^{+pH}}{\gamma_-} = K_w 10^{+pH} + \cdots, \tag{1.19}$$

(handwritten right margin: pH & pOH are independent of ↓)

$$[A^-] = \frac{CK_a'}{10^{-pH} + K_a'}, \tag{1.20}$$

$$[HA] = \frac{C(10^{-pH})}{10^{-pH} + K_a'}. \tag{1.21}$$

(handwritten: rearrange 1.15 and sub. 1.15 into 1.16)

(Construction of the diagram without numerical work is described in *Solubility and pH Calculations*, pp. 48–53 or *Ionic Equilibrium*, pp. 122–127.) The pH of the pure weak acid is the pH where the charge balance (1.17) is satisfied, or approximately

(handwritten: [OH] ≪ [H⁺] ≅ 0)

$$[H^+] = [A^-] + \cdots. \tag{1.22}$$

On the diagram, this point is given by the intersection of the $[H^+]$ and $[A^-]$ lines at pH = 3.0. (You can see from the diagram that at pH = 3.0 you get $[OH^-] = 10^{-11}$, which is indeed small compared to $[A^-] = 10^{-3}$.)

To obtain pH algebraically, substitute (1.18) and (1.20) in (1.22) and note that since pH is a full unit smaller than pK_a, 10^{-pH} is large compared to K_a' in the denominator of (1.20). This gives

(handwritten left margin: $pK_a' = -\log K_a'$ ...)

$$pH = \frac{1}{2}(pK_a' - \log C\gamma_+) + \cdots = \frac{1}{2}(pK_a^0 - \log C) + \cdots, \tag{1.23}$$

---

* Some books use K_a for the activity combination $[H^+][A^-]\gamma_+\gamma_-/[HA]\gamma_0$, but I use K_a for the concentration combination at finite ionic strength and K_a^0 for the activity combination, which is numerically equal to K_a extrapolated to zero ionic strength. A third notation is the "hybrid" constant $K_a' = K_a\gamma_+ = K_a^0\gamma_0/\gamma_-$, so that $K_a' = 10^{-pH}[A^-]/[HA]$, where $10^{-pH} = [H^+]\gamma_+$ (see pp. 34–38).

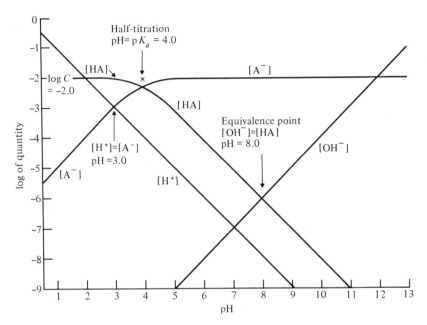

Fig. 1.1.   Logarithmic concentration diagram for a weak acid with $K_a = 10^{-4}$ and total concentration $C = 10^{-2}$.

where the approximations $\gamma_0 = 1.0$ and $\gamma_+ = \gamma_-$ were made to simplify the result. Note, however, that (1.23) applies even if γ_+ and γ_- are significantly different from 1.0, so long as they are equal.

If strong acid (with concentration C_a) is added, pH decreases, and a term in $[Cl^-]$ is added to the charge balance:

$$[H^+] = [A^-] + [OH^-] + [Cl^-]$$

or

$$[H^+] = [A^-] + [OH^-] + C_a. \qquad (1.24)$$

As soon as C_a is large compared with $[A^-]$, it dominates the right-hand side of Eq. (1.24), which becomes simply $[H^+] = C_a$. ($[OH^-]$ is negligible unless C_a and $[A^-]$ are both less than 10^{-7}.) Looking at the diagram, you can see that, because $[A^-]$ decreases as $[H^+]$ increases, this approximation is good to 1% when pH has decreased by only one unit, from pH $= 3.0$ to pH $= 2.0$.

TITRATION WITH STRONG BASE

If strong base is added to a weak acid, pH increases, and a term $[Na^+] = C_b$ is added to the charge balance:

$$[H^+] + [Na^+] = [A^-] + [OH^-] + [Cl^-]. \qquad (1.25)$$

substituted into 1.25

Combine (1.15) and (1.16) to obtain an expression for $[A^-]$, and use (1.1) to express $[OH^-]$ as a function of $[H^+]$:

$$[H^+] = \frac{CK_a}{[H^+] + K_a} + \frac{K_w}{[H^+]} + C_a - C_b. \tag{1.26}$$

Equation (1.26) can be used to calculate a titration curve (pH versus C_b, assuming $\gamma_+ = 1.0$ and $C_a = 0$), such as the one shown in Fig. 1.2.* Two inflection points may be noted on Figs. 1.1 and 1.2. The first (minimum slope) is the "half titration point," when $C_b = \frac{1}{2}C$. Provided $[H^+]$ and $[OH^-]$ are both small compared to C, Eq. (1.26) simplifies to

$$\frac{CK_a}{[H^+] + K_a} = \frac{1}{2}C + \cdots,$$

which can be further simplified to give

$$[H^+] = K_a + \cdots$$

or

$$pH = pK_a' + \cdots. \tag{1.27}$$

(marginal handwritten notes:) $10^{-2} \cdot 10^{-4}$ | $10^{-4} + 10^{-4}$ | $= 2 \times 10^{-4}$ | $\log 2 \times 10^{-4}$ | $= -2.361$ | $\big\}$

This is the point (pH = 4.0) on Fig. 1.1 where the $[HA]$ and $[A^-]$ curves cross, at a point equal to $\frac{1}{2}C$ (-2.3 log units if $C = 10^{-2}$), or 0.3 logarithmic units below the horizontal lines at log C.

The second inflection point (maximum slope on Fig. 1.2) is the *equivalence point*, where $C_b = C$. Here $[OH^-]$ can no longer be neglected, but $[H^+]$ is normally small compared with both K_a and $[OH^-]$. This assumption, applied to (1.26), yields

$$[H^+] = 0 = \frac{K_w}{[H^+]} + \frac{CK_a}{[H^+] + K_a} - C + \cdots$$

or

$$\frac{K_w}{[H^+]} = \frac{C[H^+]}{[H^+] + K_a} + \cdots. \tag{1.28}$$

Equation (1.28) is equivalent to $[OH^-] = [HA]$. If $[H^+] \ll K_a$, it reduces to

$$[H^+]^2 = \frac{K_w K_a}{C} + \cdots \tag{1.29}$$

---

* Given $C_a = 0$ and constant values of $C = 10^{-2}$ and $K_a = 10^{-4}$, choose a series of values for pH, calculate $[H^+] = 10^{-pH}$ and from Eq. (1.26) obtain

$$C_b = \frac{CK_a}{[H^+] + K_a} + \frac{K_w}{[H^+]} - [H^+].$$

If C_a is not zero, replace C_b by $C_b - C_a$. The slightly more complicated version in terms of titrant volumes is presented in *Ionic Equilibrium*, pp. 154–157 (see also Chapter 3).

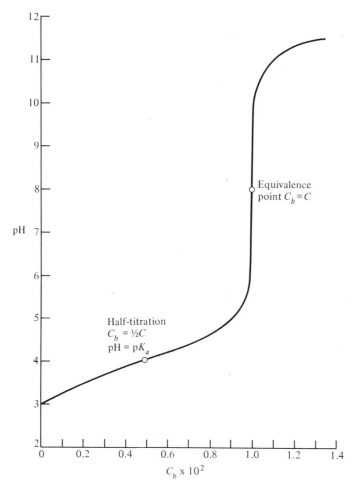

Fig. 1.2.   Titration curve of weak acid ($C = 10^{-2}$, $K_a = 10^{-4}$) with strong base ($C_b = [Na^+]$).

or*

$$pH = \frac{1}{2}(pK_w^0 + pK_a + \log C) + \cdots \tag{1.30}$$

The intersection of the $[HA]$ and $[OH^-]$ lines on Fig. 1.1 at $pH = 8.0$ gives the same result as this simplified quadratic.

---

* In transforming $[H^+]$ to pH, both sides of Eq. (1.29) must be multiplied by γ_+^2. If the assumption is made that $\gamma_+ = \gamma_-$, then $K_w \gamma_+ \gamma_- = K_w^0$. K_a remains a concentration constant.

The $[H^+]$ and $[OH^-]$ lines remain fixed on the diagram, changing only if the ionic medium or temperature changes K_w; but the $[HA]$ and $[A^-]$ lines shift if C or K_a is changed. If C is increased or decreased, the two lines move up or down together; and if K_a is increased or decreased, they shift together to the left or right so that their intersection remains at $pH = pK_a = -\log K_a$.

WEAK BASES

The only difference between a neutral weak base and the anion of a weak acid is in charge type. A weak base B (such as NH_3) can accept protons according to the equilibrium

$$BH^+ \rightleftharpoons H^+ + B$$

$$[H^+][B] = K_a[BH^+]$$

or

$$10^{-pH}[B] = K_a'[BH^+], \tag{1.31}$$

which is precisely the same form as the weak acid equilibrium (1.15). K_a is the dissociation constant of the weak acid BH^+, and the weak acid equations can be used simply by substituting BH^+ for HA, and B for A^-. Indeed, the salt Na^+A^- may be thought of as a negatively charged weak base.*

POLYPROTIC ACIDS

An acid that can yield more than one proton usually does so in steps, with the first proton being more easily removed than the second:

$$H_2A \rightleftharpoons H^+ + HA^-, \qquad HA^- \rightleftharpoons H^+ + A^-.$$

The equilibrium expressions are†

$$[H^+][HA^-] = K_{a1}[H_2A], \tag{1.32}$$

$$[H^+][A^=] = K_{a2}[HA^-]. \tag{1.33}$$

Combined, they give

$$[H^+]^2[A^=] = K_{a1}K_{a2}[H_2A]. \tag{1.34}$$

The mass balance on A is

$$C = [H_2A] + [HA^-] + [A^=]; \tag{1.35}$$

---

* The full description of these transformations, with examples, can be found in *Solubility and pH Calculations*, pp. 53–61.

† As above, hybrid constants can be used, for which $[H^+]$ is replaced by 10^{-pH} and the equilibrium constants by their primed versions: $K_1' = K_{a1}\gamma_+$, $K_2' = K_{a2}\gamma_+$, etc. (see Chapters 2 and 5).

and when you substitute the equilibrium expressions (1.32), (1.33), and (1.34) in the mass balance (1.35) to yield one equation in $[H^+]$ and $[H_2A]$, you obtain

$$[H_2A]\left(1 + \frac{K_{a1}}{[H^+]} + \frac{K_{a1}K_{a2}}{[H^+]^2}\right) = C$$

or

$$[H_2A] = \frac{C[H^+]^2}{[H^+]^2 + K_{a1}[H^+] + K_{a1}K_{a2}}. \tag{1.36}$$

Similarly,

$$[HA^-] = \frac{C[H^+]K_{a1}}{[H^+]^2 + K_{a1}[H^+] + K_{a1}K_{a2}}, \tag{1.37}$$

$$[A^=] = \frac{CK_{a1}K_{a2}}{[H^+]^2 + K_{a1}[H^+] + K_{a1}K_{a2}}. \tag{1.38}$$

The charge balance is

$$[Na^+] + [H^+] = [HA^-] + 2[A^=] + [OH^-] + [Cl^-], \tag{1.39}$$

which can be expressed in terms of $[H^+]$, by substituting from Eqs. (1.36), (1.37), and (1.38), together with the ion product of water (1.1) and the mass balances $C_b = [Na^+]$ and $C_a = [Cl^-]$:

$$[H^+] = \frac{C(K_{a1}[H^+] + 2K_{a1}K_{a2})}{[H^+]^2 + K_{a1}[H^+] + K_{a1}K_{a2}} + \frac{K_w}{[H^+]} + C_a - C_b. \tag{1.40}$$

Dissolved carbon dioxide (carbonic acid) is a diprotic acid (with a few complications resulting from its gaseous nature); the details of these equations and their applications will be developed more fully in Chapters 2–4.

BUFFERS

A weak acid or base will tend to moderate any changes in pH resulting from the addition of other acids or bases. Since the details of the buffer index are derived in Chapter 3 from first principles, you need no additional background on this topic.

SOLUBILITY PRODUCT

When a solid salt is in equilibrium with a solution containing its ions, the equilibrium expression is called the *solubility product*.* For a salt MX, forming ions M^+ and X^-,

$$[M^+][X^-] \le K_{s0}. \tag{1.41}$$

---

* As above, K_{s0} is used here for the concentration solubility product. The activity product is represented by $K_{s0}^0 = [M^+][X^-]\gamma_+\gamma_- = K_{s0}\gamma_+\gamma_-$.

If the product of these two ionic concentrations is less than K_{s0}, the salt will dissolve; if the ion product is larger than K_{s0}, salt will precipitate. The ion product can be less than K_{s0} at equilibrium only if there is no solid salt MX present.

The *common-ion effect* is the suppression of solubility by the addition of a more soluble salt containing either M^+ (i.e., MCl) or X^- (i.e., NaX) in concentration C. If S is the solubility (the number of moles per liter of the salt MX dissolved in a solution containing C moles per liter of the salt NaX), mass balances on the two ions yield

$$[M^+] = S, \tag{1.42}$$

$$[X^-] = S + C. \tag{1.43}$$

Combined with the solubility product (1.41), these mass balances (1.42) and (1.43) yield an equation formally identical with that for the pH of a strong base (with $[H^+]$ in Eq. (1.11) replaced by S, and K_w replaced by K_{s0}):

$$S(S + C) = K_{s0}. \tag{1.44}$$

Thus, for C large compared to S, the solubility is

$$S = \frac{K_{s0}}{C} + \cdots$$

and the solubility decreases with increasing C. For $C = 0$, $S = K_{s0}^{1/2}$.

The equations for more complicated salts are of course somewhat different (you will find a more complete discussion in *Solubility and pH Calculations*, Chapter 2, or *Ionic Equilibrium*, Chapter 6).

In this book, we will be primarily concerned with the solubility product of calcium carbonate ($CaCO_3$), introduced in Chapter 4, but you will encounter $Mg(OH)_2$ in Chapter 6. The solubility of this latter salt is a straightforward function of pH:

$$S = [Mg^{++}] = \frac{K_{s0}^M}{[OH^-]^2} = \frac{K_{s0}^M}{K_w^2}[H^+]^2. \tag{1.45}$$

The solubility of $CaCO_3$ is a more complicated function of pH because of the protonation of $CO_3^=$, and so I will leave that exposition until Chapter 4.

PROBLEMS*

1. Solve the three simultaneous equations

$$yz = 10^{-5}w, \qquad y = z, \qquad w + z = 10^{-3},$$

where w, y, and z are all positive real numbers.

   Answer: $z = 9.5 \cdot 10^{-5}$

---

* See also problems in *Solubility and pH Calculations*.

2. Solve the quadratic equation

$$10^3 z^2 + z = 10^{-3},$$

where z is a positive real number.

Answer: $6.2 \cdot 10^{-4}$

3. Evaluate the expression

$$C = 10^{-3.28} + \frac{10^{-5.33}}{10^{-2.09}}$$

Answer: $10^{-2.96}$

4. Evaluate the expression

$$f(I) = \frac{I^{1/2}}{1 + I^{1/2}} - 0.2I$$

for $I = 10^{-2.22}$.

Answer: $10^{-1.15}$

5. Find the pH of the following solutions at 25°C ($K_w = 10^{-14.00}$):

a) 10^{-4} M HCl   b) $10^{-4.2}$ M NaOH

c) 10^{-4} M HCl mixed with $10^{-4.2}$ M NaOH   d) $10^{-7.2}$ M HCl

Answer: (a) 4.00   (b) 9.80   (c) 4.43   (d) 6.87

6. Find the concentration of acetic acid ($pK_a = 4.75$) that will make a solution in water with pH = 4.00.

Answer: $C = 10^{-3.18}$

7. Construct the diagram of Fig. 1.1 by evaluating Eqs. (1.22) through (1.25).

8. What is the highest pH that can be achieved with boric acid ($pK_a = 8.6$) if no strong base is added?

Answer: 7.0

9. If 10^{-3} M boric acid (HB) is titrated with strong base (NaOH), find the pH at which half the boric acid is converted to borate (B^-)($pK_a = 8.6$).

Answer: 8.6

10. Evaluate Eq. (1.29) with $pK_a = 4.0$, $C = 10^{-2}$, $C_b = 10^{-1}$, $C_a = 0$ to obtain the titration curve of Fig. 1.2.

11. What pH can be achieved with 10^{-3} M NH_3 ($pK_a = 9.25$)?

Answer: 10.13

12. Use Eqs. (1.45) to (1.47) to plot a logarithmic concentration diagram analogous to Fig. 1.1. Use $pK_{a1} = 6.3$, $pK_{a2} = 10.3$.

Answer: see Fig. 2-3

13. Find the solubility of $BaSO_4$ ($pK_{s0} = 10.0$) in a sulfate solution of the same concentration as sulfate in seawater ($2.8 \cdot 10^{-2}$ M).

Answer: $3.6 \cdot 10^{-9}$

14. Find the solubility of $Mg(OH)_2$ ($pK_{s0} = 10.7$) in a solution containing $1.0 \cdot 10^{-4}$ M magnesium ion, and find the pH of the saturated solution.

Answer: $10^{-3.65}$, pH = 10.65

15. Evaluate algebraically the terms represented by "$+ \cdots$" in Eqs. (1.27) and (1.29).

First answer: $\dfrac{C}{2}(10^{-pH} - K'_w - K'_a(K'_w 10^{+pH} - 10^{-pH}))$

16. Show that if $pH = -\log([H^+]\gamma_+)$ and $\gamma_+ = \gamma_-$, the pH of pure water is $\frac{1}{2}pK^0_w$, regardless of the exact value of γ_+ or γ_-.

17. Show that the pH of a weak acid HA is given by the solution to the cubic polynomial

$$10^{-pH} = \frac{CK^0_a\gamma_0}{10^{-pH} + K'_a} + K^0_w 10^{+pH},$$

provided $\gamma_+ = \gamma_-$ (but independent of their actual values). Simplify to obtain Eq. (1.23).

18. Plot Eq. (1.26) to obtain Fig. 1.2, using $C_a = 0$, $C = 10^{-2}$, and $K_a = 10^{-4}$.

19. Show that the pH of a weak base of concentration C is given by

$$pH = \tfrac{1}{2}(pK^0_w + pK_a + \log C) \quad \text{if} \quad pH \gg pK_a.$$

Show that this is equivalent to

$$pH = \tfrac{1}{2}(pK_w + pK^0_a + \log C - \log \gamma_0).$$

The Basic Equations

The basic equations for the carbon dioxide system consist of the equilibria, a charge balance, and in some cases a mass balance on carbonate.

HENRY'S LAW

Carbon dioxide gas dissolves in water to an extent determined by its partial pressure, and by the interaction of dissolved carbon dioxide with other solutes in the water. In acid solutions (pH < 5), the concentration of carbon dioxide in solution is normally expressed by Henry's Law:

$$[CO_2] = K_H P_{CO_2}. \tag{2.1}$$

The Henry's Law constant K_H is about $10^{-1.5}$ at 25°C when $[CO_2]$ is in moles per liter and P_{CO_2} is in atmospheres. It varies slightly with the composition of the other ions in solution, as can be seen by comparing values for fresh water ($10^{-1.47}$) and seawater ($10^{-1.54}$). At higher temperatures, CO_2 is less soluble: K_H decreases to $10^{-1.7}$ at 50°C. More detailed data are given in Tables 2.1 and 2.2. The relationship of K_H to K_H^0 and ionic strength is discussed on pp. 33–34.

HYDRATION

When carbon dioxide dissolves in water, it hydrates to yield H_2CO_3. Since this reaction is slow compared to the ionization of H_2CO_3, measurements on the millisecond time scale can distinguish between simple dissolved CO_2 and hydrated species. However, at equilibrium $[H_2CO_3]$ is only about 10^{-3} as large as $[CO_2]$ and has no special significance in the acid–base equilibria, since both are uncharged. Hence I will abbreviate $[H_2CO_3] + [CO_2]$ simply as $[CO_2]$ from here on.*

---

* A number of textbooks refer to the sum $[CO_2] + [H_2CO_3]$ as $[H_2CO_3^*]$ and others refer to it simply as $[H_2CO_3]$. To me the first seems cumbersome and the second is inaccurate. The most precise notation would be $[CO_2]_T$, indicating the total of all uncharged species containing CO_2, whatever they were.

Table 2.1

Concentration equilibrium constants at 25°C and different ionic strengths*

Constant	Ionic strength						
	0	0.1	0.2	0.7	1.0	1.0	3.5
pK_H	1.47	1.48 (SW)	1.49 (SW)	1.54 (SW)		1.51 (NaClO$_4$)	1.55 (NaClO$_4$)
pK_{a1}	6.35	6.15 (KCl)	6.06 (0.26 KNO$_3$)	5.86 (SW)	5.99 (KNO$_3$)	6.04 (NaClO$_4$)	6.33 (NaClO$_4$)
pK_{a2}	10.33	9.92 (NaCl)	9.82 (NaCl)	8.95 (SW)	9.37 (NaCl)	9.57 (NaClO$_4$)	9.56 (NaClO$_4$)
pK_w	13.999	13.78 (NaClO$_4$)	13.73 (0.26 KNO$_3$)	13.20 (SW)	13.74 (NaCl)	13.78–13.86 (NaClO$_4$)	14.42 (3.8 M NaClO$_4$)
pK_{s0} (calcite)	8.34–8.52	7.16 (SW)	6.87 (SW)	6.34 (SW)		6.93 (NaClO$_4$)	7.18 (NaClO$_4$)

* $pK_H = -\log K_H$, $pK_{a1} = -\log K_{a1}$, $pK_{a2} = -\log K_{a2}$, $pK_{s0} = -\log K_{s0}$. See footnote to Table 2.2 regarding pK_{s0}^0 and ion-pairing effects. SW—seawater of appropriate salinity, i.e., $I = 0.7$ (35‰), taken from Table 5.5 (Hansson scale). See also data in Figs. 2.6 to 2.9. Data selected from Sillen, L. G. and Martell, A. E. *Stability Constants*. Special publ. no. 17 (1964) and 25 (1971). London: The Chemical Society; Riley, J. P. and Skirrow, G. 1975. *Chemical Oceanography*, vol. 2. New York: Academic Press, pp. 173–180, Butler, J. N. and Huston, R. 1975. *J. Phys. Chem.* 74:2976–2983; Edmond, J. M. and Gieskes, J. M. T. M. 1970. *Geochim. Cosmochim. Acta* 34: 1261–1291.

IONIZATION

In basic solutions, hydrated carbon dioxide ionizes to give hydrated protons $[H^+]$, bicarbonate ion $[HCO_3^-]$, and carbonate ion $[CO_3^=]$. Although the equilibrium concentration of uncharged CO_2 is not affected at constant partial pressure of CO_2, the total amount of dissolved carbonate increases with increasing pH because of the ionization equilibria.

These are represented by the equations:

$$[H^+][HCO_3^-] = K_{a1}[CO_2] \tag{2.2}$$

$$[H^+][CO_3^=] = K_{a2}[HCO_3^-]. \tag{2.3}$$

In dilute solutions at 25°C, K_{a1} is approximately $10^{-6.3}$ and K_{a2} is approximately $10^{-10.3}$. Their dependence on temperature and on the presence of other salts (Tables 2.1 and 2.2) is more complicated than for K_H, because these equilibria involve charged species. This is discussed later in more detail.

pH DEPENDENCE, CONSTANT PARTIAL PRESSURE

In a solution containing only carbon dioxide and simple acids like HCl or bases like NaOH, equations (2.1), (2.2), and (2.3) will permit you to calculate the dependence of carbonate species' concentrations on pH.

Take the logarithm of Eq. (2.1):

$$\log[CO_2] = \log K_H + \log P_{CO_2} + \cdots$$
$$= -1.5 + \log P_{CO_2} + \cdots. \tag{2.4}$$

Table 2.2
Selected values of equilibrium constants, $0°$ to $50°C$ (extrapolated to zero ionic strength)*

Temperature ($°C$)	pK_H^0 (mole/L·atm)	pK_{a1}^0 (mole/L)	pK_{a2}^0 (mole/L)	pK_w^0 (mole/L)^2	pK_{s0}^0 (Calcite) (mole/L)^{2†}	pK_{s0}^0 (corrected for ion pair)†
0	1.11	6.579	10.625	14.955	8.03	8.37
5	1.19	6.517	10.557	14.734	8.09	8.39
10	1.27	6.464	10.490	14.534	8.15	8.41
15	1.33	6.419	10.430	14.337	8.22	8.45
20	1.41	6.381	10.377	14.161	8.28	8.48
25	1.47	6.352	10.329	13.999	8.34	8.52
30	1.53	6.327	10.290	13.833	8.40	8.57
35	1.59	6.309	10.250	13.676	8.46	8.62
40	1.64	6.298	10.220	13.533	8.51	8.66
45	1.68	6.290	10.195	13.394	8.56	8.71
50	1.72	6.285	10.172	13.263	8.62	8.76
100	1.99	6.45	10.16	12.27	9.62	
150	2.07	6.73	10.33	11.64	10.54	
200	2.05	7.08	10.71	11.28	11.62	

* pK_H^0, pK_{a1}^0, pK_{a2}^0, pK_{s0}^0 were selected by Stumm, W. and Morgan, J. J. 1981. *Aquatic Chemistry* 2d ed. New York: Wiley, pp. 204–206. Some interpolations have been made to fill out the table at $5°$ intervals. Full bibliography and actual literature values (including pK_w^0) are found in Sillen, L. G. and Martell, A. E. *Stability Constants*. Special publ. no. 17 (1964) and 25 (1971). London: The Chemical Society.
$\dagger$ The first column gives the traditional solubility product, extrapolated from measurements at low ionic strength. However, the presence of ion pairs, particularly $CaHCO_3^+$, biases the extrapolation to a pK_{s0}^0 that is too low. When corrections for this ion pair are made, the values in the second column are obtained. For accurate calculations, the equilibrium $[CaHCO_3^+] = 10^{+1.23}[Ca^{++}][HCO_3^-]$ should be included (see Chapters 4 and 5). (Christ, C. L., Hostetler, P. B., and Siebert, R. M. 1974. *J. Res. US Geol. Survey* 2:175–184.)

Add to (2.4) the logarithm of Eq. (2.2), noting that $pH = -\log([H^+]\gamma_+) = -\log[H^+] + \cdots$:

$$\begin{aligned} \log[HCO_3^-] &= \log K_{a1} + \log[CO_2] + pH + \cdots \\ &= \log K_{a1} + \log K_H + \log P_{CO_2} + pH + \cdots \\ &= -7.8 + \log P_{CO_2} + pH + \cdots . \end{aligned} \tag{2.5}$$

Add to (2.5) the logarithm of Eq. (2.3):

$$\begin{aligned} \log[CO_3^=] &= \log K_{a1} + \log K_{a2} + \log[CO_2] + 2pH + \cdots \\ &= \log K_{a1} + \log K_{a2} + \log K_H + \log P_{CO_2} + 2pH + \cdots \\ &= -18.1 + \log P_{CO_2} + 2pH + \cdots . \end{aligned} \tag{2.6}$$

Note: In Eqs. (2.4), (2.5), and (2.6), the three dots indicate that approximate numerical values of K_H, K_{a1}, and K_{a2} for $25°C$ and low ionic strength are used for illustrative purposes. The numerical constants are different in other media at other temperatures (see Tables 2.1 and 2.2 and the discussion at the end of this chapter, pp. 30–40).

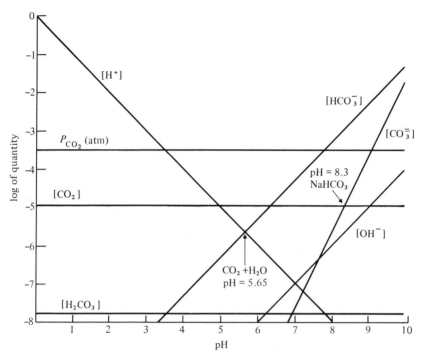

Fig. 2.1. Logarithmic concentration diagram for CO_2 at constant partial pressure of $10^{-3.5}$ atm, with added NaOH (or strong acid) to adjust pH.

EXAMPLE 1   Figure 2.1 shows this dependence for a constant partial pressure of $10^{-3.5}$ atm (the average CO_2 content of the earth's atmosphere). Note that the logarithm of each of the variables is a linear function of pH.   ∎

In addition, Eqs. (2.4), (2.5), and (2.6) tell you that the concentration of all three carbon dioxide species increases directly with the partial pressure of CO_2 in the gas phase; their ratios remain constant and independent of P_{CO_2}, as you can see by subtracting any pair of the three equations.

What pH is reached when only carbon dioxide and water are in equilibrium? Charged species are introduced because of the ionization reactions, and the number of positive charges in a given solution must be exactly balanced by the same number of negative charges:

$$[H^+] = [HCO_3^-] + 2[CO_3^=] + [OH^-]. \tag{2.7}$$

Note that, because $CO_3^=$ carries two charges, each mole dissolved must be counted twice in the charge balance.

Using the diagram in Fig. 2.1, you can easily see that for pH < 10, the value of $[HCO_3^-]$ is much larger than the other two concentrations on the right-hand side of Eq. (2.7). As a first approximation, therefore, the charge balance can be stated as

$$[H^+] = [HCO_3^-] + \cdots. \tag{2.8}$$

The answer is found where these two lines intersect, at pH $= 5.65$. Note that both $[CO_3^=]$ and $[OH^-]$ are negligible ($< 10^{-8}$) at this point. Alternatively, Eqs. (2.1) and (2.2) can be combined and substituted in (2.8) to give

$$[H^+]^2 = K_{a1}K_H P_{CO_2} + \cdots. \tag{2.9}$$

This equation, evaluated with $K_{a1} = 10^{-6.3}$, $K_H = 10^{-1.5}$, $P_{CO_2} = 10^{-3.5}$, gives pH $= 5.65$, in agreement with the graphical result.*

Increasing P_{CO_2} by a factor of 100 (two log units) decreases pH by only one unit, an example of the buffering capacity of the carbon dioxide system. Take the negative log of Eq. (2.9):

$$pH = +3.9 - \frac{1}{2} \log P_{CO_2} + \cdots.$$

The chart of Fig. 2.1, or the full charge balance (2.7), tells you that as the partial pressure of CO_2 is decreased, the concentrations of HCO_3^- and $CO_3^=$ eventually become less than that of OH^-, and the pH of the carbon dioxide solution approaches that of pure water.

THE CHARGE BALANCE IN THE PRESENCE OF BASE

With the partial pressure of CO_2 fixed, it is necessary to add base (such as NaOH) to raise the pH. Now you must include in the charge balance the ions of the dissociated base. $[OH^-]$ is already on the right side, but $[Na^+]$ must be added to the left side:

$$[H^+] + [Na^+] = [HCO_3^-] + 2[CO_3^=] + [OH^-]. \tag{2.10}$$

Since $[H^+]$ decreases rapidly as strong base is added, not much NaOH will be required before $[Na^+]$ dominates the left side of the equation.

When the amount of base added equals the total of dissolved carbon dioxide species, the solution has the stoichiometry of sodium bicarbonate (NaHCO₃):[†]

$$[Na^+] \doteq [CO_2] + [HCO_3^-] + [CO_3^=] \tag{2.11}$$

(compare with the mass balance on carbonate, Eq. (2.18) below). Combine (2.11) with (2.10) and get

$$[H^+] + [CO_2] = [CO_3^=] + [OH^-]. \tag{2.12}$$

---

* If activity coefficients are included, (2.9) becomes pH $= \frac{1}{2}(pK_{a1}^0 + pK_H^0 - \log P_{CO_2}) + \cdots$ (see Eq. 5.1).
† The base could be CaCO₃, and provided the solution is not saturated, the same equations would apply, with $[Na^+]$ replaced by $2[Ca^{++}]$ (see Chapters 4 and 5).

If both $[H^+]$ and $[OH^-]$ are small, this condition is satisfied by the intersection of the $[CO_2]$ and $[CO_3^=]$ lines on Fig. 2.1, at pH $= 8.3$:

$$[CO_2] = [CO_3^=] + \cdots . \qquad (2.13)$$

Using the equilibria (2.1), (2.2), and (2.3) in (2.13) gives

$$K_H P_{CO_2} = \frac{K_H P_{CO_2} K_{a1} K_{a2}}{[H^+]^2} + \cdots ,$$

which rearranges to

$$[H^+]^2 = K_{a1} K_{a2} + \cdots \qquad (2.14)$$

or*

$$pH = \frac{1}{2}(pK_{a1} + pK_{a2}) + \cdots . \qquad (2.15)$$

With the approximate values of K_{a1} and K_{a2}, this yields pH $= 8.3$.

As you can see from Fig. 2.1, all three terms on the right side of Eq. (2.10) increase as pH is increased, but $[CO_3^=]$ increases with a slope twice as great as that of $[HCO_3^-]$ and $[OH^-]$. Thus, as pH increases, $[CO_3^=]$ becomes more and more important compared to the singly charged species, but at constant $P_{CO_2} = 10^{-3.5}$, carbonate ion does not dominate (2.10) for pH < 10 (see Problem 6).

ALKALINITY

In the literature of natural waters, an important concept is the *alkalinity*: the hypothetical amount of strong base that must be neutralized in order to reach a pH corresponding to a solution of pure CO_2 and water.

In our simple model, consisting of only NaOH and CO_2 in water, alkalinity is given directly by Eq. (2.10) as the amount of strong base: if $[Na^+] = A$,

$$A = [HCO_3^-] + 2[CO_3^=] + [OH^-] - [H^+]. \qquad (2.16)$$

Equation (2.16) is the conceptual definition of carbonate alkalinity. The operational definition (titrate the sample with standard HCl to the methyl orange endpoint) applied to a natural water may also include other weak acids and bases, such as borate, ammonia, hydrolyzed ferric and aluminum ion, or organic acid anions, and is called total alkalinity (see Chapters 3 and 5). For the model considered here, containing only CO_2 and NaOH, total alkalinity is the same as carbonate alkalinity.

One important property of the alkalinity is that it does not change when CO_2 is added to or withdrawn from the solution. Since $A = [Na^+]$, and addition or removal of CO_2 does not affect the sodium ion concentration, it does not affect the combination of concentrations given by Eq. (2.16). For example, if the carbon

---

* If activity coefficients are included, (2.15) becomes pH $= \frac{1}{2}(pK_1' + pK_2')$.

dioxide partial pressure is changed, and $[H^+]$, $[OH^-]$, and $[CO_3^=]$ are small compared to $[HCO_3^-]$ in (2.16), the quantity

$$A = [HCO_3^-] + \cdots = \frac{K_{a1}K_H P_{CO_2}}{[H^+]} + \cdots$$

must remain constant, and therefore $[H^+]$ increases more or less linearly with increasing P_{CO_2} (see Problem 12).

ACIDITY

How can the pH of a solution be less than that of pure CO_2 and water? This is possible if a strong acid such as HCl is added. The concept of *mineral acidity** is introduced by including $[Cl^-]$ in the charge balance:

$$[Na^+] + [H^+] = [HCO_3^-] + 2[CO_3^=] + [OH^-] + [Cl^-]$$
$$A' = [Cl^-] - [Na^+] = [H^+] - [HCO_3^-] - 2[CO_3^=] - [OH^-]. \quad (2.17)$$

Comparison with Eq. (2.16) shows that mineral acidity A' is in fact just the negative of alkalinity A.

SOLUTION WITHOUT GAS PHASE

A solution contained in a vessel, so that its total carbon dioxide content is constant, describes some situations better than a solution in equilibrium with a constant partial pressure of carbon dioxide. This case has been saved for last because the equations are more complicated. The principles are the same, however, with the addition of a mass balance on carbonate:

$$C_T = [CO_3^=] + [HCO_3^-] + [CO_2] + [H_2CO_3]. \quad (2.18)$$

Note that here we are counting carbon atoms and, unlike in the charge balance, $[CO_3^=]$ is counted only once. In addition, the uncharged species are included. As pointed out earlier, $[H_2CO_3]$ is always about 0.1% of $[CO_2]$, and will therefore be omitted from subsequent equations.

Substituting Eqs. (2.2) and (2.3) in Eq. (2.18), you can get a relationship between C_T, $[H^+]$, and any one of the other concentrations. For example, if you express everything in terms of $[H^+]$ and $[CO_3^=]$, Eq. (2.3) becomes

$$[HCO_3^-] = \frac{[H^+][CO_3^=]}{K_{a2}}. \quad (2.19)$$

---

* See Stumm, W. and Morgan, J. J. 1981. *Aquatic Chemistry*, 2d ed. New York: Wiley, p. 186 for definitions of four other types of alkalinity and acidity corresponding to different titration endpoints.

(2.2) and (2.3) together give

$$[CO_2] = [H^+]^2 \frac{[CO_3^=]}{K_{a1}K_{a2}}. \tag{2.20}$$

Substitute (2.19) and (2.20) in (2.18):

$$C_T = [CO_3^=]\left(1 + \frac{[H^+]}{K_{a1}} + \frac{[H^+]^2}{K_{a1}K_{a2}}\right).$$

This equation in turn can be solved for $[CO_3^=]$ and the expression in brackets simplified to obtain:

$$[CO_3^=] = C_T \frac{K_{a1}K_{a2}}{K_{a1}K_{a2} + K_{a1}[H^+] + [H^+]^2}. \tag{2.21}$$

If you now substitute this expression in (2.18) above, you get

$$[HCO_3^-] = C_T \frac{K_{a1}[H^+]}{K_{a1}K_{a2} + K_{a1}[H^+] + [H^+]^2} \tag{2.22}$$

and similarly

$$[CO_2] = C_T \frac{[H^+]^2}{K_{a1}K_{a2} + K_{a1}[H^+] + [H^+]^2}. \tag{2.23}$$

These are general results for a diprotic acid (compare with Eqs. (1.36) to (1.38)).

A few minutes with a calculator will give you a rough table of the ratios $[CO_3^=]/C_T$, $[HCO_3^-]/C_T$, and $[CO_2]/C_T$ as functions of pH alone.*

Those functions are graphed on Fig. 2.2; it is clear that on logarithmic scales, large portions are essentially straight-line segments with short curved portions where the slope changes from $+1$ to 0 or from 0 to -1, etc. You can see and prove algebraically (see Problem 14) that the linear portions with slope $+1$ or -1 intersect the horizontal at points where $pH = pK_{a1}(-\log K_{a1})$ and $pH = pK_{a2}$. The sections with slope $+2$ and -2 intersect the horizontal and each other at the point where $pH = \frac{1}{2}(pK_{a1} + pK_{a2})$. Thus, once you know the values of K_{a1} and K_{a2}, the entire diagram can be constructed without calculations, by using only a rule for the straight portions and a steady hand for the curved portions.

When the value of C_T is changed, the shape of the diagram does not change, but the lines for the three concentrations move up and down in concert while the lines for $[H^+]$ and $[OH^-]$ remain fixed. An example is shown in Fig. 2.3, with $C_T = 10^{-3}$ M.

---

* Some texts use the notation α for the ratio of a dissociation product to the total concentration, but the subscripting is not consistent. In the notation of *Ionic Equilibrium*, or *Solubility and pH Calculations*, the subscript is the number of protons on the species: $\alpha_0 = [CO_3^=]/C_T$, $\alpha_1 = [HCO_3^-]/C_T$, and $\alpha_2 = [CO_2]/C_T$. In contrast, Stumm and Morgan (ibid.) (as well as other texts) use a notation where the subscript is the number of negative charges: $\alpha_0 = [CO_2]/C_T$, $\alpha_1 = [HCO_3^-]/C_T$, and $\alpha_2 = [CO_3^=]/C_T$. In the oceanographic literature, α is used to represent isotopic fractionation factors as well as the solubility of CO_2 (K_H in my notation). I have decided to avoid such confusion in this book by not using the symbol α at all.

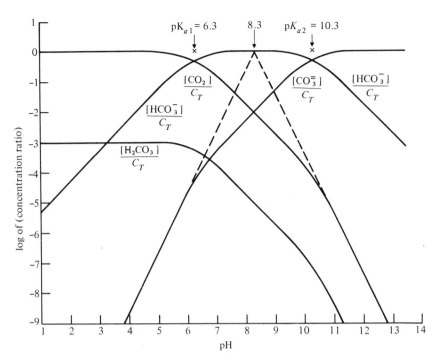

Fig. 2.2. Logarithmic concentration diagram showing the ratios of carbonate species to total carbonate as a function of pH.

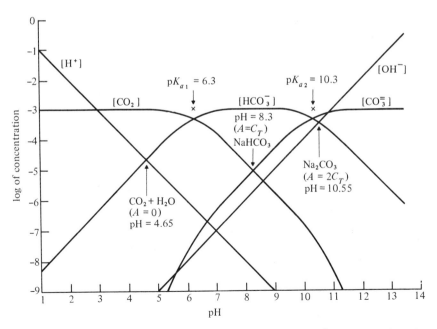

Fig. 2.3. Logarithmic concentration diagram for $C_T = 10^{-3}$ mole/L with points corresponding to $NaHCO_3$ and Na_2CO_3 noted.

CHARGE BALANCE AND PROTON CONDITION

Precisely the same charge balance (Eqs. (2.7), (2.10), or (2.17)) is used in the case where C_T is constant, as was used when P_{CO_2} was constant. The results are somewhat different, however.

EXAMPLE 2   Find the pH of $1.0 \cdot 10^{-3}$ M CO_2 in H_2O. The charge balance is Eq. (2.7):

$$[H^+] = [HCO_3^-] + 2[CO_3^=] + [OH^-] \tag{2.7}$$

and, as you can see from Fig. 2.3, the second and third terms on the right are negligible unless C_T becomes very small (then $[OH^-]$ becomes dominant instead of $[HCO_3^-]$). As before, the pH of this solution is given by the point where two lines intersect:

$$[H^+] = [HCO_3^-] + \cdots, \tag{2.8}$$

which is marked on Fig. 2.3 at pH $= 4.65$. Note that, although Eq. (2.8) still applies, the numerical result is different from that obtained in Fig. 2.1.

Knowing the approximate answer is important if you try to solve the problem algebraically. Equation (2.22) substituted in (2.8) gives:

$$K_{a1}K_{a2} + K_{a1}[H^+] + [H^+]^2 = C_T K_{a1} + \cdots,$$

but this can be simplified. Note from Fig. 2.3 that pH is in the range of 4 to 5. If pH $= 4.5$, for example, $K_{a1}K_{a2} = 10^{-16.6}$, $K_{a1}[H^+] = 10^{-10.8}$, and $[H^+]^2 = 10^{-9}$. Thus $[H^+]^2$ is much larger than the other terms, and these can be neglected to yield the approximate result

$$[H^+]^2 = C_T K_{a1} + \cdots$$

or

$$pH = \frac{1}{2}(pK_{a1} - \log C_T) + \cdots. \tag{2.24}$$

If activity coefficients are included, (2.24) is only slightly modified (see p. 36). With $pK_{a1} = 6.3$ and $C_T = 10^{-3}$, you get pH $= 4.65$, i.e., the same result as from the graphical method. Compare also with Eq. (1.23), p. 6.

Note also that Eq. (2.24) is similar to Eq. (2.9), derived with P_{CO_2} held constant. In this pH range, $[CO_2]$ is the dominant term of C_T (Eq. (2.18); also see Figs. 2.2 and 2.3); and if there were a gas phase in equilibrium with this solution, it would have a partial pressure of CO_2 given by Eq. (2.1):

$$P_{CO_2} = \frac{[CO_2]}{K_H} = \frac{C_T}{K_H} + \cdots,$$

and hence

$$[H^+]^2 = K_{a1}K_H P_{CO_2} + \cdots, \tag{2.9}$$

which is the same as was derived earlier.   ■

EXAMPLE 3   What is the pH of 10^{-3} M $NaHCO_3$? We saw above that when the solution composition corresponds to $NaHCO_3$,

$$[Na^+] = C_T = [CO_2] + [HCO_3^-] + [CO_3^=] \tag{2.11}$$

and

$$[H^+] + [CO_2] = [CO_3^=] + [OH^-]. \tag{2.12}$$

The pH of this solution is given to be 8.3 by the intersection of the $[CO_2]$ and $[CO_3^=]$ lines on Fig. 2.3 (or by Eq. (2.15)), since both $[H^+]$ and $[OH^-]$ are small compared to $[CO_2]$ and $[CO_3^=]$ at this pH.   ∎

EXAMPLE 4   What is the pH of 10^{-3} M Na_2CO_3? When the solution composition corresponds to Na_2CO_3,

$$[Na^+] = 2C_T = 2([CO_2] + [HCO_3^-] + [CO_3^=]). \tag{2.25}$$

Substitute this in (2.10) to obtain

$$[H^+] + 2[CO_2] + [HCO_3^-] = [OH^-]. \tag{2.26}$$

Note from Fig. 2.3 that when $[HCO_3^-] = [OH^-]$, both $[H^+]$ and $[CO_2]$ are negligible, so that the intersection of the two lines

$$[HCO_3^-] = [OH^-] + \cdots \tag{2.27}$$

at pH = 10.55 gives the answer.

Algebraically, substitute the ion product of water (Eq. (1.1)) and Eq. (2.22) in (2.27) to obtain

$$\frac{K_w}{[H^+]} = C_T \frac{K_{a1}[H^+]}{K_{a1}K_{a2} + K_{a1}[H^+] + [H^+]^2} + \cdots,$$

which rearranges to

$$\left(C_T \frac{K_{a1}}{K_w} - 1\right)[H^+]^2 = K_{a1}K_{a2} + K_{a1}[H^+] + \cdots.$$

Substitute $C_T = 10^{-3}$, $K_{a1} = 10^{-6.3}$, $K_{a2} = 10^{-10.3}$, $K_w = 10^{-14.0}$, and you get

$$10^{+4.70}[H^+]^2 - 10^{-6.3}[H^+] - 10^{-16.6} = 0.$$

Use the quadratic formula to obtain $[H^+] = 10^{-10.55}$, in agreement with the graphical result. This example shows how much simpler the graphical solution sometimes can be, compared to the algebraic method.   ∎

RELATION BETWEEN ALKALINITY, pH, AND C_T

When there is no CO_2-containing gas phase in equilibrium with the solution, raising the pH by addition of strong base produces more negatively charged species but does not change the total carbonate content of the solution. The results are therefore very different from the constant P_{CO_2} case, where the gaseous reservoir is called upon to increase the total carbonate in solution as the alkalinity is increased.

As before, alkalinity is defined as the concentration of strong base in the model solution

$$A = [Na^+] = [HCO_3^-] + 2[CO_3^=] + [OH^-] - [H^+] \qquad (2.16)$$

but the concentrations of HCO_3^- and $CO_3^=$ are now given in terms of C_T by Eqs. (2.21) and (2.22). Figure 2.4 shows the alkalinity function (heavy line); except for the factor of 2 on $[CO_3^=]$, it is essentially an envelope of the largest concentration of negatively charged ions over the pH range above 4.65.

From Fig. 2.4, you can see that with constant C_T, and pH between 6 and 11, alkalinity is only slightly dependent on pH, and (within a factor of 3) equal to C_T over that range.

The curve of Fig. 2.4 shows that, if C_T is fixed, a unique value of A is found for each pH. Any two of these three variables can be considered independent, and the full relationship between them is obtained by combining Eqs. (2.16), (2.21), and (2.22):*

$$A = C_T \frac{K_{a1}[H^+] + 2K_{a1}K_{a2}}{[H^+]^2 + K_{a1}[H^+] + K_{a1}K_{a2}} + \frac{K_w}{[H^+]} - [H^+]. \qquad (2.28)$$

At fixed pH, A is a linear function of C_T:

$$A = C_T F - G, \qquad (2.29)$$

where F and G are constants. This property has been used in a graphical calculation method by K. S. Deffeyes.[†]

If A and pH are given, C_T is obtained explicitly by inverting (2.29):

$$C_T = \frac{A + G}{F}. \qquad (2.30)$$

These equations will be discussed further in Chapter 3.

ACIDITY WITH CONSTANT C_T

Here there is less difference from the constant P_{CO_2} case, since for pH below zero alkalinity (pH < 4.65 in Figs. 2.3 and 2.4), essentially all carbon dioxide is present in solution as uncharged dissolved CO_2, and pH has little effect on $[CO_2]$. Therefore, mineral acidity (defined as the concentration of added strong acid, such as

---

* Refer to the beginning of the next chapter, and you will recognize Eq. (2.28) as being equivalent to Eqs. (3.6) to (3.8) with $V = 0$.

[†] *Limnol. Oceanogr.* 1965. 10:412; see also Stumm, W. and Morgan, J. J. op. cit., pp. 188–191.

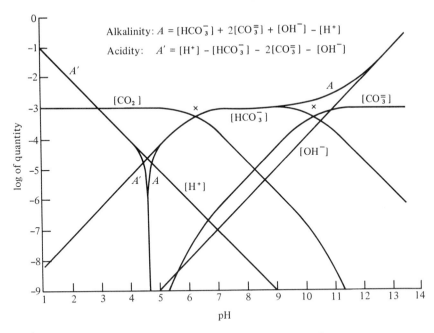

Fig. 2.4. Logarithmic concentration diagram for $C_T = 10^{-3}$ mole/L with curves for alkalinity A and mineral acidity A' superimposed.

HCl—compare with Eq. (2.17)) is simply

$$A' = [Cl^-] = [H^+] - [HCO_3^-] - 2[CO_3^=] - [OH^-]$$
$$= [H^+] - [HCO_3^-] + \cdots. \tag{2.31}$$

The last two terms are always negligible compared to the second, and for pH < 4 the first is dominant; A' is shown as a function of pH in Fig. 2.4.

Note that when pH is sufficiently low or C_T is sufficiently low, $A' = [H^+]$ and the presence of carbon dioxide has no effect on the pH of the solution (see Problem 10).

A NOTE ON CONCENTRATION UNITS

The engineering literature often expresses concentrations in terms of weight, not moles. This is awkward when a parameter such as alkalinity is involved, because it consists of a sum of several (often interconvertible) species with different formula weights:

$$A = [HCO_3^-] + 2[CO_3^=] + \cdots \quad \text{or} \quad A = [Na^+] + 2[Ca^{++}] + \cdots.$$

This problem is normally taken care of by expressing alkalinity in terms of the equivalent weight of $CaCO_3$. Every mole of $CaCO_3$ that dissolves or precipitates changes the alkalinity by two equivalent units.

Table 2.3

Chemical	Formula weight (g/mole)	Factor to obtain equivalent ppm of $CaCO_3$
$CaCO_3$	100.09	1.0000
$Ca(OH)_2$	74.09	1.3509
CaO	56.08	1.7848
$CaCl_2$	110.99	0.9018
$CaCl_2 \cdot 2H_2O$	147.02	0.6808
Ca^{++}	40.08	2.4973
CO_2	44.01	2.2743
$NaHCO_3$	84.00	0.5958
Na_2CO_3	105.99	0.9443
$NaOH$	40.00	1.2511
HCl	36.46	1.3726
H_2SO_4	98.08	1.0205

EXAMPLE 5   An alkalinity of 10^{-3} mole/L is equivalent to $0.5 \cdot 10^{-3}$ mole/L of $CaCO_3$. Since the formula weight of $CaCO_3$ is

$$40.08 + 12.01 + 3 \cdot 16.00 = 100.09 \text{ g/mole},$$

this concentration may be expressed as

$$(100 \text{ g/mole}) \cdot 0.5 \cdot 10^{-3} \text{ mole/L} = 50 \cdot 10^{-3} \text{ g/L} = 50 \text{ mg/L}$$

or 50 parts per million by weight (ppm) of $CaCO_3$.   ■

Any quantity that is related to alkalinity or calcium ion concentration can be expressed in terms of ppm $CaCO_3$ by using simple stoichiometry to determine how many moles of $CaCO_3$ would be required to produce the same (or, in the case of acids, the opposite) effect on the alkalinity. Table 2.3 gives factors for converting weight concentrations of a number of common reagents used in adjusting the composition of natural waters or wastewaters.

EXAMPLE 6   One mole of $CaCl_2$, hydrate (such as $CaCl_2 \cdot 2H_2O$), or Ca^{++} in solution is equivalent to one mole of $CaCO_3$, but for each of these compounds the formula weight is different: 110.99, 147.02, and 40.08 g/mole, respectively. Consequently, the factor by which the concentration of each chemical (in mg/L) is multiplied to convert to ppm of $CaCO_3$ is different in each case:

$$\frac{100.09}{110.99} = 0.9018, \qquad \frac{100.09}{147.02} = 0.6808, \qquad \frac{100.09}{40.08} = 2.4973. \qquad \blacksquare$$

EXAMPLE 7 One mole of $NaHCO_3$ changes alkalinity by one equivalent unit, and hence is equivalent to $\frac{1}{2}$ mole of $CaCO_3$. Therefore, the factor for converting ppm $NaHCO_3$ to ppm $CaCO_3$ is

$$\frac{100.09}{2 \cdot 84.00} = 0.5958. \qquad \blacksquare$$

Hardness is usually defined as the sum of calcium and magnesium ion concentrations in solution, and if these concentrations are expressed in moles per liter there is no problem. However, since both are often determined by a single titration with a chelating agent (such as ethylene diamine tetraacetic acid, or EDTA), it is conventional in the engineering literature to express hardness as ppm $CaCO_3$. This means that the number of equivalents of chelating agent required for the titration is converted to ppm with the factor 100.09 g/mole. The actual combined weight of $CaCO_3$ and $MgCO_3$ if both were precipitated would be less because of the lower atomic weight of Mg.

EXAMPLE 8 Water is analyzed and reported to have a total hardness (Mg and Ca) of "205 ppm as $CaCO_3$." This means

$$[Ca^{++}] + [Mg^{++}] = \frac{205 \cdot 10^{-3} \text{ g/L}}{100 \text{ g/mole}} = 2.05 \cdot 10^{-3} \text{ mole/L}.$$

Additional analysis shows that the calcium hardness is 182 ppm. This means that $[Ca^{++}] = 1.82 \cdot 10^{-3}$ mole/L. The difference between total hardness and calcium hardness is a measure of the magnesium concentration, but you must be careful to use the formula weight of $CaCO_3$, *not* the formula weight of Mg^{++} or $MgCO_3$, in converting from ppm to moles per liter:

$$\text{Mg hardness} = 205 - 182 \text{ ppm} = 23 \text{ ppm as } CaCO_3.$$

$$\text{Correct:} \quad [Mg^{++}] = \frac{23 \cdot 10^{-3} \text{ g/L}}{100 \text{ g/mole}} = 0.23 \cdot 10^{-3} \text{ mole/L};$$

$$\text{Incorrect:} \quad [Mg^{++}] \neq \frac{23 \cdot 10^{-3} \text{ g/L}}{24.3 \text{ g/mole}};$$

$$\text{Incorrect:} \quad [Mg^{++}] \neq \frac{23 \cdot 10^{-3} \text{ g/L}}{84.3 \text{ g/mole}}. \qquad \blacksquare$$

IONIC-STRENGTH EFFECTS

In most of the examples of Chapters 1–4, I emphasize how pH is related to alkalinity when carbonate species are present by using equilibrium constants for 25°C and dilute media. However, you have already seen from Table 2.1 that these "constants" change somewhat when other salts are added to the solution, even when there are no direct chemical reactions between the added salts and the carbonate species.

The derivation of even the most elementary theory of ionic interactions (Debye and Hückel's original model of ions as electrostatically charged hard spheres) is beyond the scope of this book; and the more subtle improvements on this theory are even more complicated. However, the results of the simpler semiempirical theory are easy to use, and I will present them without formal derivation.*

One important concept and mathematical form that comes out of the Debye–Hückel theory is the ionic strength, the combination of concentrations that is the primary determinant of the activity coefficient of charged species:

$$I = \frac{1}{2} \sum_i C_i z_i^2, \qquad (2.32)$$

where C_i is the concentration of an ion with charge z_i.

EXAMPLE 9   For a simple 1–1 electrolyte such as sodium chloride, the ionic strength is

$$I = \frac{1}{2}([Na^+] + [Cl^-]) = C,$$

where C is the concentration of the salt.   ■

EXAMPLE 10   For a mixture of NaCl at concentration C_1 and Na_2SO_4 at concentration C_2, the ionic strength is

$$I = \frac{1}{2}([Na^+] + [Cl^-] + 4[SO_4^=])$$

$$= \frac{1}{2}(C_1 + 2C_2 + C_1 + 4C_2) = C_1 + 3C_2.   ■$$

---

* For an elementary discussion of activity coefficients, see *Ionic Equilibrium*, Chapter 12. The most recent compendium of research results can be found in Pytkowicz, R. M., ed. 1979. *Activity Coefficients in Electrolyte Solutions*. Boca Raton, Florida: CRC Press. Three classic books are: Lewis, G. N. and Randall, J. T. 1923, revised 1961. *Thermodynamics*. New York: McGraw-Hill; Harned, H. S. and Owen, B. B. 1958. *Physical Chemistry of Electrolytic Solutions*, ed. 3. New York: Reinhold; Robinson, R. A. and Stokes, R. H. 1965. *Electrolyte Solutions*. London: Butterworths.

EXAMPLE 11   For the mixture of ionic species found at an intermediate point in the titration of CO_2 with NaOH,

$$I = \frac{1}{2}([Na^+] + [H^+] + [HCO_3^-] + [OH^-] + 4[CO_3^=]).$$

Note that as the titration proceeds, the ionic strength increases.   ■

The activity of a single ionic species cannot be directly measured without some empirical assumption (see below), but combinations of activity coefficients corresponding to neutral salts are accessible to rigorous thermodynamic measurements. For 1–1 electrolytes, the parameter measured by electrochemical or isopiestic (vapor-pressure) methods is

$$\gamma_\pm = (\gamma_+\gamma_-)^{1/2}. \tag{2.33}$$

For 1–2 electrolytes, such as Na_2SO_4,

$$\gamma_\pm = (\gamma_+^2\gamma_=)^{1/3}. \tag{2.34}$$

This mean activity coefficient is plotted for three 1–1 salts and two 1–2 salts in Fig. 2.5.

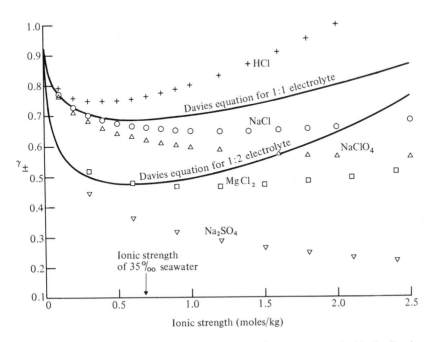

Fig. 2.5.   Activity coefficient data for 1:1 and 1:2 salts compared with the Davies equation (2.36).

Note that the differences between salts of the same charge type are small compared to the primary effect of ionic strength until the ionic strength exceeds about 0.1. You can imagine that the introduction of one or more adjustable parameters greatly increases the accuracy of predicting activity coefficients at higher ionic strengths for solutions containing only one electrolyte.

Solutions containing one or two electrolytes have been the subject of considerable study, and for many simple mixtures accurate empirical results are available.* However, few data are available for mixtures containing several cations and several anions, and the amount of data required to cover fully even the common cases encountered in this book would be enormous.

One important simplification results if the major ionic components do not change concentration very much as the reactions proceed to equilibrium and the ionic strength is therefore nearly constant. For example, the first ionization constant of CO_2 at finite ionic strength (K_{a1}) is related to the extrapolated zero ionic strength constant (K_{a1}^0) by a combination of activity coefficients:

$$K_{a1} = \frac{[H^+][HCO_3^-]}{[CO_2]} = K_{a1}^0 \frac{\gamma_0}{\gamma_+ \gamma_-}, \qquad (2.35)$$

where γ_0, γ_+, and γ_- are all constant if the ionic strength of the medium does not change. Since K_{a1}^0 depends only on temperature and pressure, then if the activity coefficients are held constant, K_{a1} will depend on temperature, pressure, and ionic medium, but *not* on pH or total carbon dioxide concentration.[†]

In Table 2.1, values of equilibrium constants in several different ionic media are given. In the laboratory, it is common to use a medium of 0.5 to 3.5 M sodium perchlorate, a salt whose ions have only weak interactions with most other ions. Seawater is the classic natural water of constant ionic strength (although more complex in composition than a simple salt) and is discussed in more detail in Chapter 5.

THE DAVIES EQUATION

In dilute solutions, I prefer to use the Davies equation, which has no adjustable parameters, even though it may be less accurate in some cases than more complicated empirical expressions.[‡] For an ion with charge z, the activity coefficient

* R. M. Pytkowicz, ibid.; R. A. Robinson and R. H. Stokes, ibid.
† In a constant ionic medium, the standard state to which thermodynamic data are extrapolated is not zero concentration of all ions, but rather a finite concentration of certain ions (e.g., Na^+ and ClO_4^-) and zero concentration of all other ions. The equilibrium constants measured in constant ionic media are therefore just as "thermodynamically" well defined as those constants extrapolated to infinite dilution.
‡ A survey of simple empirical equations for γ and their comparison with equilibrium data is given by Sun, M. S., Harriss, D. K., and Magnuson, V. R. 1980. *Can. J. Chem.* 58:125–1257.

is given by

$$\log \gamma = -0.5z^2 f(I)$$

$$f(I) = \left(\frac{I^{1/2}}{1 + I^{1/2}} - 0.2I\right)\left(\frac{298}{t + 273}\right)^{2/3}. \qquad (2.36)$$

Here t is temperature in degrees centigrade, and the function applies only to the temperature range 0–$50°C$. For accurate work the temperature dependence of the dielectric constant of water should also be taken into account. (See Pytkowicz *ibid.*, or Robinson and Stokes, *ibid.*)

The mean activity coefficient (compare Eq. (2.33) or (2.34)) is given by

$$\log \gamma_\pm = -0.5z_+z_- f(I)$$

(see Problem 23); and this function is plotted on Fig 2.5 for comparison with the experimental data given there. At ionic strength above 0.5 M, the experimentally determined activity coefficients for many salts are significantly different from those predicted by the Davies equation.

Deviation to lower values can be partially accounted for by introducing an ion-pairing equilibrium to the chemical model. For example, Na_2SO_4 solutions are considered to contain $NaSO_4^-$ as well as Na^+ and $SO_4^=$. This type of model will be discussed further in Chapters 4 and 5. It is particularly applicable to sea-water and brines.

EFFECT OF IONIC STRENGTH ON K_H

The activity coefficient of uncharged species (such as CO_2 and NH_3) is usually close to unity, and has only a small dependence on ionic strength. For example, Table 2.1 shows that pK_H is not quite independent of ionic strength, increasing from 1.47 to 1.51 at $25°C$ as the medium changes from infinite dilution to 1.0 M $NaClO_4$.* Since neither K_H^0 nor P_{CO_2} is affected by the ionic strength of the solution, this change is ascribed to a change in γ_0 (compare Eq. (2.1)):

$$K_H^0 = \frac{[CO_2]\gamma_0}{P} = K_H\gamma_0. \qquad (2.37)$$

By definition, $\gamma_0 = 1.0$ when $I = 0$. Taking logarithms of both sides of (2.37) gives

$$\log \gamma_0 + \log K_H = \log K_H^0,$$

$$\log \gamma_0 = pK_H - pK_H^0 = bI. \qquad (2.38)$$

Using the $NaClO_4$ data, you find that $b = +0.04$, which is not typical. If the data in Table 2.1 for seawater are used, the effect is somewhat larger, with a coefficient

---

* Nilsson, G., Rengemo, T., and Sillén, L. G. 1958. *Acta Chem. Scand.* 12:868. See also Sillén, L. G. and Martell, A. E. 1964. *Stability Constants.* (# 17) London: The Chemical Society, pp. 134–135.

Table 2.4

Ionic strength dependence of Henry's law constant for
CO_2 in NaCl solutions*

Temperature	b
10	0.11
30	0.10
50	0.09
176	0.08
265	0.13
330	0.20

* $\log \gamma_0 = bI$ for $I = 0$ to 2.0 molal NaCl. Data presented here
are from A. J. Ellis and R. M. Golding, 1963, *Am. J. Sci.* 261:
47–60, and S. D. Malinin, *Geokhimiya*, 1959, No. 3, pp 235–245;
see Fig. 5.24.

An empirical equation for the temperature (t) and ionic
strength (I) dependence of the activity coefficient of dissolved CO_2
was presented by Wigley, T. M. L. and Plummer, L. N., 1976.
Geochimica et Cosmochimica Acta 40:989–995, based on data for
the solubility of CO_2 in NaCl solutions obtained by Harned,
H. S., and Davis, R. Jr., 1943. *J. Am. Chem. Soc.* 65:2030–2037:

$$\log \gamma_0 = \frac{(33.5 - 0.109t + 0.0014t^2)I - (1.5 + 0.015t + 0.004t^2)I^2}{t + 273}$$

of $+0.10$ instead of $+0.04$. Table 2.4 gives values of b for NaCl solutions over a
range of temperatures. These values are not too different from the linear term in
the Davies equation ($+0.1I$) or from those obtained for other uncharged molecules
like NH_3.*

EFFECT OF IONIC STRENGTH ON FIRST ACIDITY CONSTANT K_{a1}

The important equilibrium constants for the carbonate system are easily expressed
as functions of ionic strength (for $I < 1.0$) by using the Davies equation. For
many examples in Chapters 5 and 6, I have chosen to use "hybrid" equilibrium
constants instead of "concentration" constants (see p. 6) for convenience, since
pH, not $[H^+]$, is the quantity usually measured in practice. The hybrid constant is
defined by

$$K_1' = \frac{10^{-pH}[HCO_3^-]}{[CO_2]} = K_{a1}^0 \frac{\gamma_0}{\gamma_-}. \tag{2.39}$$

* J. Bjerrum gives $b = 0.12$ for NH_3 in *Metal Ammine Formation in Aqueous Solution.* 1957.
Copenhagen: P. Haase & Son. See also Sillén and Martell, op. cit., p. 150.

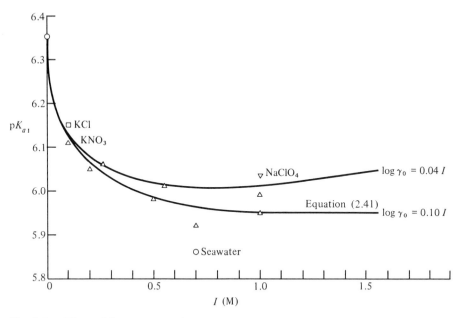

Fig. 2.6. First acidity constant of CO_2, pK_{a1}, as a function of ionic strength. Note that seawater falls well below the data for $NaClO_4$ and KNO_3 media, and that $b = 0.04$ or $b = 0.10$ give equally good fits to the data for $I \leq 1$ M.

Substituting (2.36) and (2.38) in (2.39), you get

$$pK_1' = pK_{a1}^0 + \log\gamma_- - \log\gamma_0 = pK_{a1}^0 - 0.5f(I) - bI. \qquad (2.40)$$

A selection of values for pK_{a1}^0 is given in Table 2.2 for the temperature range 0–50°C.

To compare the calculated constants with experimental data, substitute the Davies equation (2.36) in the expression for the concentration constant (2.35):

$$pK_{a1} = pK_{a1}^0 + 2\log\gamma_- - \log\gamma_0 = pK_{a1}^0 - 1.0f(I) - bI. \qquad (2.41)$$

The ionic strength dependence of pK_{a1} is graphed in Fig. 2.6. Note that Eq. (2.41) agrees with empirical data for noncomplexing supporting electrolytes ($NaClO_4$, KNO_3, KCl) when $I < 1.0$. Note also that the coefficient b in the last term of (2.41), which derives from the expression (2.38) for γ_0, could be larger than 0.10 without introducing significant error.

EXAMPLE 12   Find the pH of a 0.5 M NaCl solution containing $1.0 \cdot 10^{-3}$ M $NaHCO_3$ and saturated with CO_2 at 25°C and 0.10 atm partial pressure. The charge balance is the same as on p. 19, and the same approximation of neglecting

$[OH^-]$ and $[CO_3^=]$ applies:

$$[Na^+] + [H^+] = [HCO_3^-] + [Cl^-] + \cdots. \tag{2.17'}$$

Note that $[Na^+] - [Cl^-] = 0.5 + 1.0 \cdot 10^{-3} - 0.5 = 1.0 \cdot 10^{-3}$. However, the equilibrium constants are no longer the values for zero ionic strength, and activity coefficients must be introduced:

$$10^{-pH}[HCO_3^-]\gamma_- = K_{a1}^0[CO_2]\gamma_0 \tag{2.39'}$$

$$\gamma_0[CO_2] = K_H^0 P. \tag{2.37'}$$

By substituting (2.37') in (2.39') and using the result in (2.17') above, you can get

$$10^{-pH}\gamma_-([H^+] + 10^{-3}) = K_{a1}^0 K_H^0 P.$$

The Davies equation (2.36) gives $\gamma_- = 10^{-0.157}$ at $I = 0.5$. With $pK_{a1}^0 = 6.352$, $pK_H^0 = 1.47$, $P = 10^{-1}$, you get pH = 5.66. Note that $[H^+]$ is only 0.2% of the bracketed term and can be neglected, since it is small compared to 10^{-3}. ∎

EXAMPLE 13   Find the pH of a 0.5 M NaCl solution containing CO_2 but no alkali at a total concentration $C_T = 10^{-2}$ M. As above,

$$[H^+] = [HCO_3^-] + \cdots, \tag{2.8'}$$

$$10^{-pH}[HCO_3^-] = K_1'[CO_2] = K_1' C_T + \cdots. \tag{2.39''}$$

Expressing K_1' by using the Davies equation, as in (2.40), noting from (2.8) that $[HCO_3^-] = 10^{-pH}/\gamma_+$ and according to (2.36), $\log \gamma_+ = -0.5 f(I)$, you can get

$$pH = \frac{1}{2}(pK_{a1}^0 - \log C_T - bI)$$

(compare with Eq. (2.24), derived earlier on p. 24. For the values given above, pH = 4.151. If the ionic-strength correction were neglected, you would have calculated pH = 4.176. Note that the principal activity coefficient contributions cancel out when $[HCO_3^-] = [H^+]$, an important simplification for the equivalence point of the alkalinity titration discussed in the next chapter. ∎

EFFECT OF IONIC STRENGTH ON SECOND ACIDITY CONSTANT K_{a2}

The second acidity constant is treated analogously. The hybrid constant is given by

$$K_2' = \frac{10^{-pH}[CO_3^=]}{[HCO_3^-]} = K_{a2}^0 \frac{\gamma_-}{\gamma_=}. \tag{2.42}$$

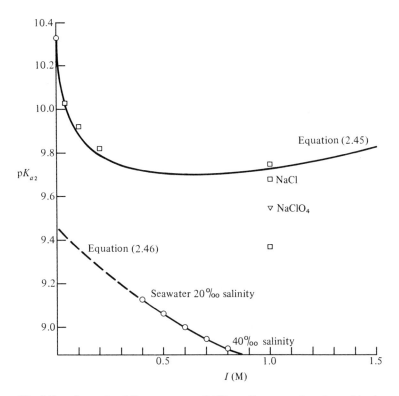

Fig. 2.7.   Second acidity constant of CO_2, pK_{a2}, as a function of ionic strength. Note the large difference between NaCl or $NaClO_4$ media (which are predicted by the Davies equation) and the observed data for seawater. This is due to ion-pair formation in seawater (see Table 5.7).

If you use the Davies equation to calculate γ_- and $\gamma_=$, you get

$$pK_2' = pK_{a2}^0 + \log\gamma_= - \log\gamma_- = pK_{a2}^0 - 1.5f(I). \qquad (2.43)$$

Values for pK_{a2}^0 at temperatures from 0 to 50°C are presented in Table 2.2.
    The concentration constant is

$$K_{a2} = \frac{[H^+][CO_3^=]}{[HCO_3^-]} = K_{a2}^0 \frac{\gamma_-}{\gamma_+\gamma_=}. \qquad (2.44)$$

With the approximation that the activity coefficients of H^+ and HCO_3^- are equal, γ_+ and γ_- in (2.44) cancel, giving

$$pK_{a2} = pK_{a2}^0 + 2\log\gamma_= = pK_{a2}^0 - 2.0\,f(I). \qquad (2.45)$$

This curve is presented graphically in Fig. 2.7 and compared with empirical data, some of which was already given in Table 2.1.

Note that the Davies equation does not give a steep enough dependence on ionic strength to fit the data for seawater. This is because of the relatively strong ion pairing of Mg^{++}, Ca^{++}, and Na^{+} with $CO_3^{=}$ and $SO_4^{=}$ (see Table 5.3). I have already mentioned one way to accommodate this effect—to introduce additional equilibria; but one can also use an empirical function with a graph steeper than the one given by the Davies equation: Hansson's data for 25°C, plotted on Fig. 2.7, are fitted by

$$pK_{a2} = 9.46 - 1.935 \cdot 10^{-2}S + 1.35 \cdot 10^{-4}S^2, \tag{2.46}$$

where S is salinity in ‰ (parts per thousand). Note that the constant in (2.46) is *not* pK_{a2}^0, and that this function is not meant to predict pK_{a2} at salinities lower than 20‰ ($I = 0.4$).

EFFECT OF IONIC STRENGTH ON ION PRODUCT OF WATER

For the ion product of water, an appropriate hybrid equilibrium constant is

$$K_w' = 10^{-pH}[OH^-] = \frac{K_w^0}{\gamma_-}; \tag{2.47}$$

when combined with the Davies equation, it gives

$$pK_w' = pK_w^0 + \log\gamma_- = pK_w^0 - 0.5f(I). \tag{2.48}$$

To compare with empirical data, the concentration constant is needed:

$$pK_w = pK_w^0 - 1.0f(I), \tag{2.49}$$

and in Fig. 2.8 this function is plotted together with data for pK_w obtained in $NaClO_4$, $NaCl$, and KNO_3 media. Nearly all these data fall higher than the curve of Eq. (2.49). This may be attributed to the large ionic size (hydration) of $H^+(aq)$. In seawater, in contrast, $pK_w = 13.2$, which is off the bottom of the graph of Fig. 2.8. This low value results from OH^- ion pairs with Mg^{++} and Ca^{++}.

EFFECT OF IONIC STRENGTH ON SOLUBILITY PRODUCT OF CaCO₃

Although the use of the solubility product of $CaCO_3$ will not be discussed in detail until Chapter 4, I will describe here how the activity coefficient concepts apply to that equilibrium also. Since $[H^+]$ does not appear explicitly in the solubility product expression, there is no hybrid equilibrium constant, only the concentration constant K_{s0} and the activity constant K_{s0}^0 (K_{s0} extrapolated to

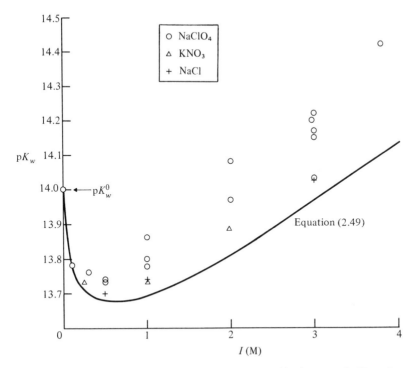

Fig. 2.8. Ion product of water, pK_w, as a function of ionic strength. Note that the Davies equation predicts lower values of pK_w than are actually observed in NaClO$_4$, KNO$_3$, and NaCl media, but that the difference is not larger than the difference between various experiments.

zero ionic strength). These are related by the activity coefficients of Ca^{++} and CO$_3^{=}$:

$$K_{s0} = [\text{Ca}^{++}][\text{CO}_3^{=}] = \frac{K_{s0}^0}{\gamma_{++}\gamma_{=}}. \tag{2.50}$$

When the activity coefficients are evaluated with the Davies equation,

$$pK_{s0} = pK_{s0}^0 + 2\log\gamma_{++} = pK_{s0}^0 - 4.0f(I). \tag{2.51}$$

Values of pK_{s0}^0 are given as a function of temperature from 0 to 50°C in Table 2.2. Two sets are given: the traditional values and a recently computed set that takes into account the CaHCO$_3^+$ ion-pair formation (this is discussed in Chapter 4).

In Fig. 2.9, the calculated values from the Davies equation are compared with empirical data. The agreement is satisfactory in NaClO$_4$, but not in seawater. Of all the equilibrium constants discussed in this book, the solubility product of calcium carbonate is the most sensitive to ionic strength and also the most sensitive to ion pairing (these models are described in more detail in Chapters 4 and 5).

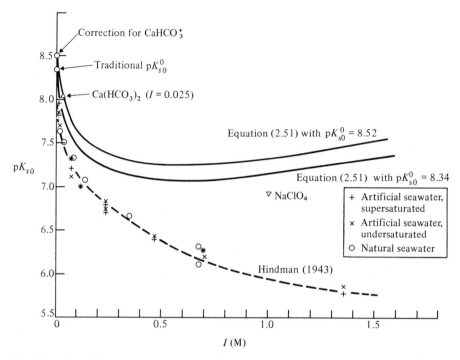

Fig. 2.9.   Solubility product of calcite, pK_{s0}, as a function of ionic strength. Note that the data for seawater are well below those for noncomplexing electrolytes, but that the Davies equation predicts the value for $NaClO_4$ medium moderately well.

Although surface waters and laboratory analyses are rarely subjected to conditions that fall outside atmospheric pressure and the temperature range of 0 to 50°C, the pressure in deep ocean waters exceeds 1000 atm and the conditions in hydrothermal solutions often exceed 300°C and many thousands of atmospheres. These cases are discussed in Chapter 5, and the appropriate data for high temperatures and pressures are presented there.

PROBLEMS

1. What would be the pH of a soda water made by saturating distilled water with pure CO_2 at 1 atm pressure?

    Answer: 3.9

2. Construct a diagram like the one in Fig. 2.1 for $P_{CO_2} = 0.1$ atm. (This might be appropriate for modeling soil pore water since metabolism by soil microorganisms produces substantially more CO_2 than is found in the average atmosphere.)

3. Find the amount of strong base required to change the pH of water saturated with CO$_2$ at 0.1 atm from its initial value of pH $= 4.4$ to pH $= 5.5$, assuming equilibrium is maintained with the gas phase.

   Answer: $C_b = 10^{-3.3}$ M

4. What is the alkalinity of a solution containing 10^{-2} M NaHCO$_3$ after it has reached an equilibrium with an atmosphere containing CO$_2$ at partial pressure of 10^{-2} atm?

   Answer: 10^{-2} M

5. How much NaOH must be added to water saturated with CO$_2$ at 10^{-2} M partial pressure to achieve the solution composition corresponding to NaHCO$_3$?

   Answer: $[Na^+] = 10^{-1.49}$ M

6. If NaOH is added to a water saturated with CO$_2$ at $10^{-3.5}$ atm and the solution is kept in equilibrium with the atmosphere, at what pH will the composition of Na$_2$CO$_3$ be reached?

   Answer: The proton condition $[HCO_3^-] + \cdots = [OH^-]$ is never satisfied because both these concentrations increase with pH at the same slope (see Fig. 2.1).

7. If water is saturated with CO$_2$ at 10^{-2} M partial pressure and then the vessel is sealed so that no more CO$_2$ can be absorbed and NaOH is added, how much NaOH is required to reach the composition corresponding to NaHCO$_3$?

   Answer: $[Na^+] = C_T = 10^{-3.48}$ (compare with Problem 5).

8. Use Eqs. (2.21) to (2.23) to draw a diagram similar to the ones in Figs. 2.2 and 2.3.

9. A solution containing 10^{-3} M NaHCO$_3$ is treated with excess strong acid so that pH is reduced to 3.5. What is the partial pressure of CO$_2$ in equilibrium with this solution?

   Answer: $10^{-1.5}$ atm

10. A solution containing 10^{-2} M Na$_2$CO$_3$ is treated with excess strong acid so that the final pH is 2.0. What is the acidity of this final solution?

    Answer: $A' = 10^{-2.0}$ M

11. A solution containing 10^{-3} M NaHCO$_3$ is treated with $5 \cdot 10^{-4}$ M Na$_2$CO$_3$. What is the alkalinity of the final solution?

    Answer: $2 \cdot 10^{-3}$ M

12. In a natural water there are many additional cations and anions, including Na$^+$, Ca^{++}, Mg^{++}, K$^+$, Cl$^-$, SO$_4^=$, SiO(OH)$_3^-$, etc. Prove that the alkalinity $A = [HCO_3^-] + 2[CO_3^=] + [OH^-] - [H^+]$ of a water containing the above-listed species does not change when CO$_2$ is added or withdrawn. Note: pK_a for HSO$_4^-$ is between 1.1 and 2.0, and pK_a for Si(OH)$_4$ is between 9.2 and 9.6.

13. Derive an equation giving the variation of pH with partial pressure of CO$_2$ at constant alkalinity $A = 2 \cdot 10^{-3}$. This is a model for the effect of increased atmospheric CO$_2$ on the pH of seawater. Make the appropriate approximations for pH near 8.0.

    Answer: pH $= 5.1 - \log P + \cdots$

14. Show that the portions of the curves of log $[CO_2]$ versus pH and log $[CO_3^=]$ versus pH (in Fig. 2.2 or 2.3) with slope -1 or $+1$, when extrapolated to C_T, intersect the horizontal at pK_{a1} and pK_{a2}. Show that these lines have slope -2 and $+2$ for sufficiently low values

of $[CO_2]$ or $[CO_3^=]$ and that the portions of these curves with slope $+2$ or -2, when extrapolated to the concentration C_T, intersect at $pH = \frac{1}{2}(pK_{a1} + pK_{a2})$.

15. Find C_T for $pH = 8.0$ and $A = 2 \cdot 10^{-3}$. Simplify Eq. (2.37) by dropping negligible terms.

    Answer: $C_T = 2.03 \cdot 10^{-3}$

16. Express alkalinity of $2 \cdot 10^{-3}$ equivalents/L in terms of ppm as $CaCO_3$.

    Answer: 100 ppm

17. Express an HCl concentration of $1.0 \cdot 10^{-4}$ mole/L as ppm $CaCO_3$.

    Answer: 5 ppm

18. A water is analyzed and reported to have $pH = 7.2$, total hardness 175 ppm as $CaCO_3$, calcium hardness 160 ppm as $CaCO_3$, and alkalinity 165 ppm as $CaCO_3$. Find the molar concentration of $[Ca^{++}]$, $[Mg^{++}]$, $[HCO_3^-]$, $[CO_3^=]$ and speculate about the remaining ions in the water if the total dissolved solids are 280 mg/L.

    Answer: $[Ca^{++}] = 1.60 \cdot 10^{-3}$ M, $[Mg^{++}] = 1.5 \cdot 10^{-4}$ M, $[HCO_3^-] = 3.30 \cdot 10^{-3}$ M, $[CO_3^=] = 2.5 \cdot 10^{-6}$ M, $[Cl^-] = 1.95 \cdot 10^{-4}$ M.

19. Lowenthal and Marais (1976. *Carbonate Chemistry of Aquatic Systems*. Ann Arbor, Mich.: Ann Arbor Science Publishers, pp. 146–147) use a consistent set of equations for equilibria and mass and charge balances, where the concentrations (marked with asterisks) are expressed in ppm $CaCO_3$ instead of moles per liter. Show that the following equations are consistent with those derived in this chapter and find numerical values for the constants K_H^*, K_{a1}^*, K_{a2}^*, K_w^*, and K_{s0}^*:

$$A^* = [HCO_3^-]^* + [CO_3^=]^* + [OH^-]^* - [H^+]^*,$$

$$C_T^* = \frac{1}{2}[CO_2]^* + [HCO_3^-]^* + \frac{1}{2}[CO_3^=]^*,$$

$$[CO_2]^* = K_H^* P_{CO_2},$$
$$[H^+]^*[HCO_3^-]^* = K_{a1}^*[CO_2]^*,$$
$$[H^+]^*[CO_3^=]^* = K_{a2}^*[HCO_3^-]^*,$$
$$[H^+]^*[OH^-]^* = K_w^*,$$
$$[Ca^{++}]^*[CO_3^=]^* = K_{s0}^*.$$

    Answer: $K_H^* = 5 \cdot 10^4 K_H$, $K_{a1}^* = 2.5 \cdot 10^4 K_{a1}$, $K_{a2}^* = 10^5 K_{a2}$, $K_w^* = 2.5 \cdot 10^9 K_w$, $K_{s0}^* = 10^{10} K_{s0}$.

20. *Acid rain* is produced when acidic air pollutants (H_2SO_4 and HNO_3) are washed out of the atmosphere by rain. Compare the pH to be expected for rain in equilibrium with a clean atmosphere ($P_{CO_2} = 10^{-3.5}$ atm) and that expected for rain that contains 10^{-5} M HNO_3 and $2 \cdot 10^{-5}$ M H_2SO_4.

    Answer: 5.65; 4.3

21. Compare the activity coefficient of Na^+ in

    a) 0.4 M NaCl   b) 0.1 M NaCl + 0.1 M Na_2SO_4

    Answer: The same, according to the Davies equation.

22. Calculate pK_H, pK_1', pK_2', pK_{a1}, pK_{a2}, pK_w, and pK_{s0} at $I = 1.0$ and $25°C$ by using Eqs. (2.41) through (2.49) and data in Table 2.3. Compare with data in Table 2.1.

23. The mean activity coefficient of an electrolyte is defined by

$$\gamma_\pm = (\gamma_1^{\nu_1}\gamma_2^{\nu_2})^{1/(\nu_1+\nu_2)},$$

where γ_1 is the activity coefficient of the cation, γ_2 is the activity coefficient of the anion, ν_1 is the number of moles of cations formed in solution from one formula weight of salt, and ν_2 is the number of moles of anion. Show that if Eq. (2.36) holds, this definition leads to

$$\log\gamma_\pm = -0.5z_+z_- f(I).$$

24. Compare the activity coefficient of H^+ in water saturated with CO_2 at 1 atm partial pressure and in that same solution (isolated from the gas phase) after it has been titrated with NaOH to the $NaHCO_3$ end point.

   Answer: 0.988, 0.823

25. Find the pH of 1.0 M $NaClO_4$ containing 10^{-3} M $NaHCO_3$ and saturated with CO_2 at 0.10 atm, making use of the empirical concentration constants found in Table 2.1. Compare with the similar example in the text (p. 36), where the Davies equation was used to calculate activity coefficients. In both cases an assumption about activity coefficients is needed to obtain pH. Which approach would be expected to yield the most accurate results?

   Answer: 5.70, using the Davies equation to calculate γ_+ only. The accuracy is about the same but could be improved by using a pH scale calibrated in 1.0 M $NaClO_4$.

26. Recent measurements of the meltwater from high-altitude snow in Greenland and the Himalayas have given values near pH = 5.1 instead of the expected value. Use data in Table 2.2 to calculate the pH of water at 0°C in equilibrium with CO_2 at a partial pressure of $10^{-3.5}$ atm. What would the partial pressure of CO_2 have to be in order to reduce pH to 5.1? What concentration of strong acid would be required for the same reduction in pH at $P = 10^{-3.5}$ atm?

   Answer: pH = 5.60, $10^{-2.5}$ atm, $A' = 10^{-5.15}$

The Alkalinity
Titration Curve

In Chapter 2, alkalinity was defined in terms of a model containing only CO_2 and NaOH in solution:

$$A = [HCO_3^-] + 2[CO_3^-] + [OH^-] - [H^+]. \qquad \text{(2.16) or (3.1)}$$

If such a solution is titrated with HCl or other strong acid to the pH (approximately 4.5) corresponding to pure CO_2 and H_2O, the amount of acid added will be equivalent to A, or A_c, the carbonate alkalinity.

In a natural water, other acids and bases may contribute to the reaction with HCl and produce a titer that is usually greater than the carbonate alkalinity. This empirical measure is called *total alkalinity* A_T. Before I go into details of how various additional components may affect the total alkalinity and the shape of the titration curve, I want to derive the shape of the titration curve for a solution containing only NaOH and CO_2. The equation governing this curve is closely related to Eq. (2.28) p. 26.

THE TITRATION CURVE EQUATION

Let V_0 be the initial volume of the sample, which contains alkalinity A and total carbonate C_T (these two values will determine the initial pH, so it cannot be specified in addition). Let the titrant be a strong acid (HCl) of concentration C_a, and the volume be V. Then the various concentrations in the charge balance (Eq. (2.17)) can be specified in terms of the parameters listed above and the variable $[H^+]$:

$$[Na^+] + [H^+] = [Cl^-] + [HCO_3^-] + 2[CO_3^=] + [OH^-], \qquad \text{(2.17) or (3.2)}$$

with

$$[Na^+] = \frac{AV_0}{V + V_0}, \qquad (3.3)$$

$$[Cl^-] = \frac{C_aV}{V + V_0}, \qquad (3.4)$$

$$[OH^-] = \frac{K_w}{[H^+]}. \tag{3.5}$$

$[HCO_3^-]$ and $[CO_3^=]$ are given by Eqs. (2.21) and (2.22) with C_T multiplied by $V_0/(V + V_0)$ to correct for dilution. Substitute these together with (3.3), (3.4), and (3.5) in (3.2). The result is:

$$A \frac{V_0}{V + V_0} + [H^+] = \frac{C_a V}{V + V_0} + \frac{C_T V_0}{V + V_0} \cdot \frac{K_{a1}[H^+] + 2K_{a1}K_{a2}}{K_{a1}K_{a2} + K_{a1}[H^+] + [H^+]^2} + \frac{K_w}{[H^+]}.$$

Inversion of this equation to give V as a function of $[H^+]$ is straightforward and yields

$$V = V_0 \frac{A - C_T F + G}{C_a - G}, \tag{3.6}$$

where

$$F = \frac{[HCO_3^-] + 2[CO_3^=]}{C_T} = \frac{K_{a1}[H^+] + 2K_{a1}K_{a2}}{K_{a1}K_{a2} + K_{a1}[H^+] + [H^+]^2} \tag{3.7}$$

and

$$G = [H^+] - [OH^-] = [H^+] - \frac{K_w}{[H^+]}. \tag{3.8}$$

Note the similarity to Eq. (2.28).

A little work with a calculator can produce a curve of pH versus V for any set of parameters (see Fig. 3.1). Note that the functions F and G (Fig. 3.2) are dependent only on the equilibrium constants and pH, and can be stored preprogrammed if you have that capability; G is normally small or negligible compared to A, C_a, or C_T.

An alternative form of F, useful for some purposes, is:*

$$F = \frac{K_{a1}}{[H^+] + K_{a1}} + \frac{K_{a2}}{[H^+] + K_{a2}} + \cdots, \tag{3.9}$$

which may be remembered by thinking of F as originating with CO_2—HCO_3^- and HCO_3^-—$CO_3^=$ as two independent acid–base pairs (except for the missing factor of 2, which multiplies $[CO_3^=]$ in the definition of F).

Multiplying out Eq. (3.9) into a single fraction, you obtain

$$F = \frac{K_{a1}[H^+] + K_{a1}K_{a2} + K_{a2}[H^+] + K_{a1}K_{a2}}{[H^+]^2 + K_{a1}[H^+] + K_{a2}[H^+] + K_{a1}K_{a2}}. \tag{3.10}$$

---

* As before, the notation "$+ \cdots$" indicates that an approximation (usually by neglecting one or more terms in the mass or charge balances) has been made. The additional quantity may be positive or negative, but under the conditions specified is normally less than 1% of the terms given explicitly. A valuable exercise for the reader (and not always trivial!) is to evaluate numerically the quantity represented by "$+ \cdots$".

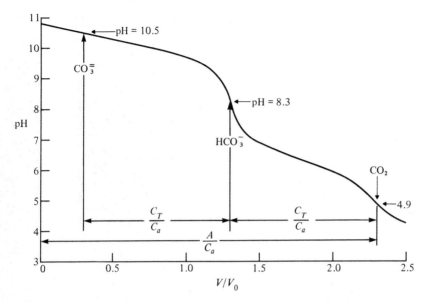

Fig. 3.1. Titration of volume V_0 of a solution ($A = 2.3 \cdot 10^{-3}$ M, $C_T = 1.0 \cdot 10^{-3}$ M) with volume V of strong acid (with concentration $C_a = 1.0 \cdot 10^{-3}$ M). Equilibrium constants: $pK'_a = 6.3$, $pK_{a2} = 10.3$, $pK_w = 14.0$.

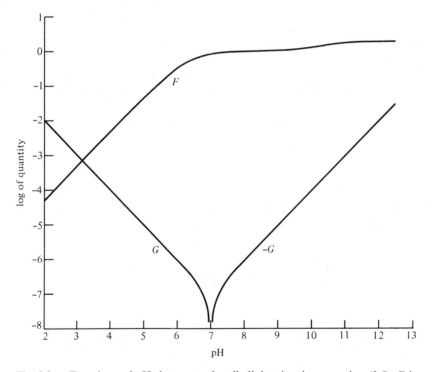

Fig. 3.2. Functions of pH that enter the alkalinity titration equation (3.6); F is given by (3.7), (3.9), or (3.11); G is given by (3.8) or (3.12).

46

Since $K_{a2} = 10^{-10.3}$ is always a factor of 10^4 smaller than $K_{a1} = 10^{-6.3}$, the term $K_{a2}[H^+]$ is negligible compared to $K_{a1}[H^+]$ for all values of pH. Thus Eq. (3.10) reduces to (3.7), given above. Note that (3.9) is a general result for all diprotic acids for which $K_{a2} \ll K_{a1}$.

Inclusion of activity coefficients is straightforward. For example, if the expressions are to be given in terms of pH $= -\log([H^+]\gamma_+)$ instead of $-\log[H^+]$, you can show (see Problem 18) that

$$F = \frac{10^{-(pH + pK_1')} + 10^{+0.30 - (pK_1' + pK_2')}}{10^{-2pH} + 10^{-(pH + pK_1')} + 10^{-(pK_1' + pK_2')}}, \tag{3.11}$$

$$G = \frac{10^{-pH} - 10^{-pK_w' + pH}}{\gamma_+}, \tag{3.12}$$

where (see p. 34)

$$K_1' = 10^{-pH} \frac{[HCO_3^-]}{[CO_2]} = K_{a1}\gamma_+ = K_{a1}^0 \frac{\gamma_0}{\gamma_-}, \tag{2.39}$$

$$K_2' = 10^{-pH} \frac{[CO_3^=]}{[HCO_3^-]} = K_{a2}\gamma_+ = K_{a2}^0 \frac{\gamma_-}{\gamma_=}. \tag{2.42}$$

These hybrid constants are particularly useful in calculations on seawater.

TITRATION ERROR

The end point of the alkalinity titration is wherever the analyst decides to end it by means of indicator or pH meter; the equivalence point V_2 corresponds to complete neutralization of the carbonate base by the strong acid:

$$[Cl^-] = [Na^+]$$

or

$$C_a V_2 = A V_0. \tag{3.13}$$

The full Eq. (3.6) contains extra terms for $[OH^-]$ and $[CO_3^=]$, which are normally negligible in the pH range (4 to 5) of interest (see Fig. 2.4). Note that in the pH range near 4.5, Eq. (3.7) or (3.9) becomes approximately

$$F = \frac{K_{a1}[H^+] + \cdots}{\cdots + [H^+]^2} = \frac{K_{a1}}{[H^+]} + \cdots \tag{3.14}$$

(verify that the other terms are small by substituting numerical values: pH $= 4.5$, $pK_{a1} = 6.3$, $pK_{a2} = 10.3$). Similarly, Eq. (3.8) becomes

$$G = [H^+] + \cdots \tag{3.15}$$

(you cannot set $G = 0$ because at the equivalence point $A = 0$, the value of $-C_T F$ must be balanced by G).

Substitute (3.14) and (3.15) in (3.6) to get

$$V = V_0 \frac{A - (C_T K_{a1}/[H^+]) + [H^+]}{C_a - [H^+]} + \cdots,$$

which rearranges to

$$C_a V = A V_0 - \frac{C_T V_0 K_{a1}}{[H^+]} + (V + V_0)[H^+] + \cdots. \tag{3.16}$$

Knowing the pH at the equivalence point, you can use (3.16) to calculate the absolute titration error: the number of equivalents of acid required to reach the end point from the equivalence point, $C_a(V - V_2)$. Make use of (3.13) and (3.16) to get

$$C_a(V - V_2) = C_a V - A V_0 = (V + V_0)[H^+] - \frac{C_T V_0 K_{a1}}{[H^+]}. \tag{3.17}$$

This equation requires that V be estimated to evaluate the right-hand side; since V is approximately V_2, Eq. (3.13) tells you to substitute $V = A V_0/C_a$:

$$C_a(V - V_2) = \frac{A + C_a}{C_a} V_0[H^+] - \frac{C_T V_0 K_{a1}}{[H^+]} + \cdots. \tag{3.18}$$

The relative titration error E is obtained by dividing the absolute titration error, as given by Eq. (3.18), by the total number $A V_0$ of equivalents of alkalinity:

$$E = \frac{C_a(V - V_2)}{A V_0} = [H^+] \frac{A + C_a}{A C_a} - \frac{K_{a1}}{[H^+]} \frac{C_T}{A} + \cdots. \tag{3.19}$$

Note that the equivalence point $V = V_2$ is reached when $E = 0$, or

$$[H^+]^2 = K_{a1} C_T \frac{C_a}{A + C_a} + \cdots, \tag{3.20}$$

which is the same as we derived earlier (Eq. 2.24) for CO_2 and H_2O alone) except for the dilution factor $C_a/(A + C_a)$.* The estimate of the pH of the equivalence point depends on a good estimate of K_{a1} as well as C_T and A, and these are not always known accurately before the titration is begun.

EXAMPLE 1   A numerical example illustrates best how the titration error is calculated. With $A = C_T = 10^{-3}$, $C_a = 10^{-1}$, and $K_{a1} = 10^{-6.3}$, the equivalence point ($E = 0$) is found from (3.20) to be pH = 4.65. Suppose the end point were set at pH = 4.8 instead. This could happen if the pH meter were incorrectly calibrated or if an indicator were used under conditions where the color change

---

* Activity coefficients are introduced on p. 49, following the example.

became visible at pH $= 4.8$. Then, from (3.19), the relative error would be

$$E = \frac{(10^{-4.80})(10^{-3} + 10^{-1})}{(10^{-3})(10^{-1})} - \frac{(10^{-6.3})(10^{-3})}{(10^{-4.8})(10^{-3})}$$

$$= 10^{-1.80} - 10^{-1.50} = -0.016.$$

The true equivalence point would be underestimated by 1.6%.   ■

EFFECT OF IONIC STRENGTH

How does ionic strength affect the equivalence point of the alkalinity titration? If Eq. (3.20) is multiplied by γ_+^2, you get

$$pH = \frac{1}{2}\left(pK_1' - \log \gamma_+ - \log \frac{C_T C_a}{A + C_a} \right) + \cdots \tag{3.21}$$

(note that $K_{a1}\gamma_+ = K_1'$). If the approximation $\gamma_- = \gamma_+$ is made in order to transform $K_{a1}\gamma_+^2$ into $K_{a1}^0\gamma_0$, you can obtain a second alternative form:

$$pH = \frac{1}{2}\left(pK_{a1}^0 - \log \gamma_0 - \log \frac{C_T C_a}{A + C_a} \right) + \cdots . \tag{3.22}$$

Equation (3.22) shows clearly the weak dependence of the equivalence point pH on ionic strength in noncomplexing media. Recall that $\log \gamma_0 = +0.1I$ (pp. 33–34) and that pK_{a1}^0 is independent of ionic strength. Thus an increase in ionic strength from 0 to 1.0 M decreases the equivalence-point pH by 0.05 units. Such a result would apply in a solution consisting mostly of NaCl, KNO_3, or $NaClO_4$.

EXAMPLE 2   However, this simple result does not apply if there is strong ion pairing, such as found in seawater. If you begin with $pK_a^0 = 6.352$ (Table 2.2) and calculate activity coefficients from the Davies equation with $I = 0.7$ (Eq. (2.40), p. 35), you obtain $pK_1' = 6.16$ instead of the experimental value 6.00 (see Table 5.5, p. 122). Similarly, using the Davies equation you get $pK_{a1} = 6.00$ instead of the experimental value 5.86 (Table 5.5).

Since $K_{a1}\gamma_+ = K_1'$, you can obtain an experimental value for γ_+ from the two independent experimental equilibrium constants:

$$\log \gamma_+ = pK_{a1} - pK_1' = 5.86 - 6.00 = -0.14. \tag{3.23}$$

This is not too different from the value -0.16 predicted by the Davies equation. As you will see in Chapter 5, the hydrogen ion is not much affected by ion pairing, but the activity of HCO_3^- is substantially reduced.

How does this affect the use of Eqs. (3.20), (3.21), and (3.22)? You can calculate pH at the equivalence point, using the three types of constants in three ways. First, from (3.20) with $pK_{a1} = 5.86$, $A = C_T = 10^{-3}$, $C_a = 10^{-1}$, you get

$$[H^+] = \left[\frac{(10^{-5.86})(10^{-3})(10^{-1})}{10^{-3} + 10^{-1}}\right]^{1/2} = 10^{-4.43}.$$

Using (3.23), you get

$$pH = -\log[H^+]\gamma_+ = 4.43 + 0.14 = 4.57.$$

The second way is to use $pK_1' = 6.00$ in Eq. (3.21):

$$pH = \frac{1}{2}(6.00 + 0.14 + 3.00) = 4.57.$$

This agrees with the first answer. Note that both these results rely entirely on experimental equilibrium-constant data and the only assumption about activity coefficients is expressed by Eq. (3.23).

On the other hand, if $pK_{a1}^0 = 6.352$ is used in (3.22), you get $pH = 4.66$. This is the least accurate of the three results, because Eq. (3.22) was derived by assuming $\gamma_+ = \gamma_-$, and this does not take into consideration the ion pairing we know occurs in seawater, whereas the constants determined experimentally in actual seawater do account for these effects. ■

The effect of ionic strength, alkalinity, and total carbonate on titration error is most easily seen from Eq. (3.19). If you prefer, you can substitute 10^{-pH} for $[H^+]\gamma_+$ to obtain the alternative form

$$E = \frac{10^{-pH}}{\gamma_+} \cdot \frac{A + C_a}{AC_a} - K_1' 10^{+pH} \frac{C_T}{A} + \cdots. \tag{3.24}$$

As above, with $\gamma_+ = \gamma_-$, you get

$$E = \frac{10^{-pH}}{\gamma_+} \cdot \frac{A + C_a}{AC_a} - \frac{K_{a1}^0 \gamma_0 10^{+pH}}{\gamma_+} \frac{C_T}{A} + \cdots. \tag{3.25}$$

Equation (3.25) shows most clearly that the magnitude of E is inversely proportional to γ_+, all other things being equal. This factor $1/\gamma_+$ varies from 1.0 at low ionic strength to a maximum of about $10^{+0.16}$ at $I = 0.7$ (compare Fig. 2.5), and, if neglected, produces a maximum effect on E of about 40% of E. The effect of γ_0 is smaller (10%) and is important only if E is negative. If there are important ion-pairing effects, these will also affect the second term of (3.25), but not (3.19) or (3.24).

Since the titrating acid concentration C_a is normally large compared to the alkalinity, the term $A + C_a$ is usually approximately C_a, and the dilution factor in the first term of (3.19), (3.24), or (3.25) is approximately $1/A$. If the end point is more acidic than the equivalence point (first term dominates and E is positive), E will vary inversely with A but be independent of C_T and nearly independent of C_a. If the end point is more basic than the equivalence point (the second term dominates and E is negative), E still varies inversely with A, but also varies directly with C_T, so that a constant ratio C_T/A will produce the same error regardless of dilution.

Thus for the same error in pH, the error in the titration E will be least for larger values of A. If C_T/A is large compared to unity or there is significant but unquantified ion pairing (i.e., K_{a1}^0 is known but no experimental values of K_{a1} or K_1' are available), less error will occur from overshooting the equivalence point (positive E) than from undershooting it. In such cases, the Gran titration (see below) becomes an important tool.

TOTAL ALKALINITY AND CARBONATE ALKALINITY

At the beginning of this chapter I mentioned the contribution made by acids and bases other than CO_2 and its anions to the total alkalinity as measured by titration with HCl. The general relationship between carbonate alkalinity and total alkalinity can be derived by considering a model solution containing CO_2, NaOH, a weak acid HA (normally as its anion A^-), and a weak base B (as its cation BH^+). In seawater, the most important HA would be HSO_4^- and the most important B would be $B(OH)_4^-$ or NH_3, but in general there might be several of each, as well as other inert ions, such as Ca^{++}, Mg^{++}, Br^-, etc.

In the initial solution (superscript zero) the charge balance is

$$[H^+]^0 + [Na^+]^0 + [BH^+]^0 = [OH^-]^0 + [HCO_3^-]^0 + 2[CO_3^=]^0 + [A^-]^0. \tag{3.26}$$

The carbonate alkalinity is given by

$$A_c = [HCO_3^-]^0 + 2[CO_3^=]^0 + [OH^-]^0 - [H^+]^0, \tag{3.1}$$

so that

$$[Na^+]^0 = A_c + [A^-]^0 - [BH^+]^0. \tag{3.27}$$

In the solution at the equivalence point (superscript prime), the charge balance is analogous to (3.26):

$$[H^+]' + [Na^+]' + [BH^+]' = [OH^-]' + [HCO_3^-]' + 2[CO_3^=]' + [A^-]' + [Cl^-]'. \tag{3.28}$$

By definition, the carbonate alkalinity is zero at the equivalence point and $[Cl^-]$, corrected for dilution, is the total alkalinity A_T:

$$[Na^+]' = [A^-]' - [BH^+]' + A_T \frac{V_0}{V + V_0}. \tag{3.29}$$

As long as Na^+ (or the combination of inert cations and anions that it represents) does not change (except for dilution) on the addition of HCl, then

$$[Na^+]^0 \frac{V_0}{V + V_0} = [Na^+]'$$

and

$$A_T = A_c + [A^-]^0 - [A^-]' \frac{V + V_0}{V_0} - [BH^+]^0 + [BH^+]' \frac{V + V_0}{V_0}. \quad (3.30)$$

The usual mass balances and equilibria (Eqs. (1.15) and (1.16)) for monoprotic acids and bases give expressions for the ionic concentrations in terms of K_a and $[H^+]$:

$$[A^-]^0 - [A^-]' \frac{V + V_0}{V_0} = \frac{C_A K_a}{K_a + [H^+]^0} - \frac{C_A K_a}{K_a + [H^+]'}. \quad (3.31)$$

If pK_a is of the order of 5 or less and the initial pH^0 is of the order of 6 to 8, then $[H^+]^0 \ll K_a$ and the first term of (3.31) is just C_A. Further simplification leads to

$$[A^-]^0 - [A^-]' \frac{V + V_0}{V_0} = \frac{C_A [H^+]'}{K_a + [H^+]'} = [HA]' + \cdots . \quad (3.32)$$

A similar transformation on the BH^+ terms in (3.30) (assuming that pK_a' for this acid–base pair is significantly greater than 5, so that $K_a' \ll [H^+]'$) yields:

$$-[BH^+]^0 + [BH^+]' \frac{V + V_0}{V_0} = [B]^0 + \cdots . \quad (3.33)$$

Then (3.32) and (3.33) can be substituted in (3.30) to give

$$A_T = A_c + [HA]' + [B]^0 + \cdots . \quad (3.34)$$

The general result for any number of weak acids (fully dissociated at the start of the titration) and weak bases (fully protonated at the equivalence point of the titration) is simply a sum of terms like (3.34):

$$A_T = A_c + \sum_i [HA_i]' + \sum_i [B_i]^0 + \cdots . \quad (3.35)$$

LINEARIZED (GRAN) TITRATION CURVES

Titration to a predetermined pH value, whether a pH meter or an indicator is used, is subject to errors if the solutions are not completely pure. A number of workers, especially in Scandinavia, have adopted a method of end point determination that depends on linearizing a portion of the titration curve and extrapolating it to zero.*

---

* Gran, G. 1952. *Analyst* 77:661–671.

The easiest way to see the principle of the Gran titration is in the context of a strong acid–base titration. Assume that a volume V_0 of strong base [NaOH] (with concentration C_0) is titrated with a volume V of strong acid [HCl] (with concentration C). The charge balance then gives the titration curve directly, by the methods used above:

$$[H^+] + [Na^+] = [Cl^-] + [OH^-],$$

$$[H^+] + \frac{C_0 V_0}{V + V_0} = \frac{CV}{V + V_0} + [OH^-]. \tag{1.8}$$

On the acid side of the titration curve (pH < 7), after the equivalence point is reached, $[OH^-]$ is negligible compared to the other concentrations, and the equation can be transformed to give a function linear in V:

$$(V + V_0)[H^+] = CV - C_0 V_0 + \cdots$$

or

$$f_1 = (V + V_0)10^{-pH} = (CV - C_0 V_0)\gamma_+ + \cdots. \tag{3.36}$$

If the experimenter determines pH as a function of V for the acid range of the titration curve (pH < 7) and plots f_1 as a function of V, he/she should obtain a straight line with slope $df_1/dV = C\gamma_+$ intersecting the horizontal axis ($f_1 = 0$) at the equivalence point $V_e = C_0 V_0/C$ (Fig. 3.3). Normally, the ionic strength does not change significantly in this part of the titration curve, and γ_+ is constant (see the discussion below and Problem 20).

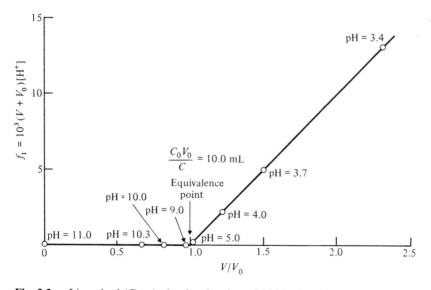

Fig. 3.3.   Linearized (Gran) plot for titration of 10.00 mL of strong base ($C_0 = 10^{-3}$ M) with strong acid ($C_a = 10^{-3}$ M).

LINEARIZING THE ACID PORTION OF
THE ALKALINITY TITRATION

By analogy with the titration of a strong base with a strong acid, the portion of any titration when excess strong acid is present can be linearized by using Eq. (3.36). For example, when a carbonate-containing solution is titrated with strong acid, the portion of the titration curve with pH < 5 is determined primarily by $[H^+]$.

Recall that the full equation for the titration curve is given by

$$V = V_0 \frac{A - C_T F + G}{C_a - G}, \tag{3.6}$$

and for pH < 5, approximations (3.14) and (3.15) apply, so that

$$V(C_a - [H^+]) = V_0 \left(A - \frac{C_T K_{a1}}{[H^+]} + [H^+] \right) + \cdots . \tag{3.37}$$

This is a familiar result from the previous discussion of titration errors (Eq. (3.16) rearranged).

When $[H^+]$ is sufficiently large, the second term of (3.37) (with C_T) becomes negligible, and the equation for this most acid portion of the titration curve is

$$V(C_a - [H^+]) = V_0(A + [H^+]) + \cdots \tag{3.38}$$

or

$$f_1 = (V + V_0)[H^+] = C_a V - A V_0 + \cdots , \tag{3.39}$$

which is the same form as Eq. (3.36) derived above for the strong acid–strong base titration.

Thus by plotting f_1 versus V in the excess acid region you can extrapolate back to the CO_2 end point ($V_2 = AV_0/C_a$) at $f_1 = 0$. This is illustrated in Fig. 3.4. Note that a substantial excess of acid must be added to get a good linear plot, because the term $C_T K_{a1}/[H^+]$ becomes significant near the equivalence point and causes much more curvature than does the $[OH^-]$ term neglected in Eq. (3.36) (compare Figs. 3.3 and 3.4).

In the illustration, the end point is at $V_2/V_0 = 2.3$ and the plot extends to $V/V_0 = 3.1$. If the plot were extended only to 2.8 (dotted line), the extrapolation might underestimate the end point by 1% ($V/V_0 = 2.27$). Similarly, if A is extremely small and a large excess of strong acid is added, γ_+ may not be perfectly constant and $[H^+]$ should be calaculated as $10^{-pH}/\gamma_+$ for each point (see Problem 21).

An alternative is to note that by Eq. (3.39) the slope df_1/dV is simply C_a, the concentration of titrant.* In Fig. 3.4 the solid line is drawn with this slope $(1.0 \cdot 10^{-3})$ through the highest points. Without further adjustment, it passes through the equivalence point at $f_1 = 0$.

---

* $C_a \gamma_+$, if $[H^+]$ is replaced by 10^{-pH}. In Fig. 3.4, where $I \cong 10^{-3}$, $\gamma_+ \cong 10^{-0.02}$, and the slope would be $10^{-3.02}$ instead of $10^{-3.00}$. For simplicity of presentation, I have used $[H^+]$ instead of 10^{-pH} in most equations.

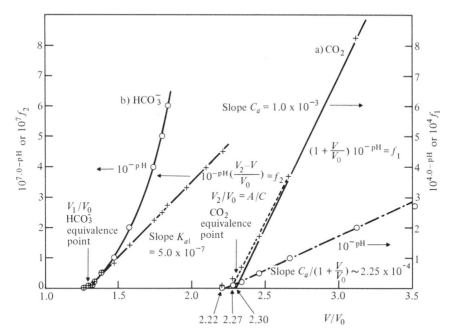

Fig. 3.4. Examples of Gran plots for the titration shown in Fig. 3.1: (a) pH range below 4.5 to determine the CO_2 equivalence point at $V_2 = V_0A/C_a$; (b) pH range 4.5 to 8.3 to determine the HCO$_3^-$ equivalence point at $V_1 = V_0(A - C_T)/C_a$.

   The circles in Fig. 3.4 represent a plot of $[H^+]$ without the factor $(V + V_0)$. The parameters used for the illustration (concentration of titrant C_a is equal to C_T) cause this factor to change by 40% over the range plotted. Hence an extrapolation of several of the circles, even though they appear to fall approximately on a straight line, will lead to an erroneous estimate of the end point, in this case 2.22, i.e., too low by about 3%.

   In the more common case where the titrant is concentrated compared to the sample, V is small compared to V_0, and the factor $V + V_0$ does not change so much. For example, if I had chosen $C_a = 10^{-1}$ instead of 10^{-3}, the V/V_0 scale would be 100 times smaller, the factor $V + V_0$ would be constant within a few percent over the entire titration, and a plot of $f_1 = [H^+]$ versus V could be extrapolated to V_2 at $f_1 = 0$ with 0.1% accuracy.

LINEARIZING THE HCO$_3^-$ REGION

A common method for measuring both C_T and alkalinity is to determine the HCO$_3^-$ equivalence point (pH $\simeq$ 8.3) as well as the CO_2 equivalence point. If $A > C_T$, the initial pH is higher than 8.3, and both the HCO$_3^-$ (V_1) and $CO_2(V_2)$

equivalence points can be reached by titrating with acid (see Fig. 3.1). If $A < C_T$, the initial pH is between 8.3 and 5, so that titration with acid will allow you to reach only the CO_2 equivalence point. To reach the HCO_3^- equivalence point, base must be added, corresponding to a negative value for V.*

For simplicity, assume that $A > C_T$, as in Fig. 3.1, so that all values of V are positive. The linear function to be developed by analogy with Eq. (3.39) will use CO_2 as the acidic species instead of H^+, and will measure the degree of conversion of HCO_3^- to CO_2 as the titration progresses from the HCO_3^- equivalence point (pH $\simeq$ 8.3)

$$V_1 = (A - C_T)\frac{V_0}{C_a} \tag{3.40}$$

to the CO_2 equivalence point (pH $\simeq$ 5)

$$V_2 = \frac{AV_0}{C_a}. \tag{3.13}$$

In making approximations to Eq. (3.5), note that in the pH range 5 to 8.3, G is negligible and F is close to 1.0 (see Fig. 3.2). However, if you simply set $F = 1.0$, you lose all the pH dependence and get only the expression (3.40) for V_1.

It is convenient to substitute $A = C_a V_2/V_0$ from (3.13) and $C_T = C_a(V_2 - V_1)/V_0$ from (3.40) combined with (3.13) in (3.6):

$$V = V_0\frac{A - C_T F}{C_a} = V_2 - (V_2 - V_1)F.$$

Add VF to both sides and rearrange:

$$(V_2 - V)(1 - F) = (V - V_1)F. \tag{3.41}$$

Now evaluate the ratio $(1 - F)/F$ from the full expression (3.7),

$$\frac{1 - F}{F} = \frac{[H^+]^2 - K_{a1}K_{a2}}{K_{a1}[H^+] + 2K_{a1}K_{a2}}. \tag{3.42}$$

Further approximation is possible. In the pH range 5 to 8.3, $[H^+]^2$ varies from 10^{-10} to $10^{-16.6}$, $K_{a1}[H^+]$ varies from $10^{-11.3}$ to $10^{-14.6}$, while $K_{a1}K_{a2}$ remains constant at $10^{-16.6}$. Thus over almost all the range, $[H^+]^2$ is the large term in the numerator of (3.42); over the whole range $K_{a1}[H^+]$ is the largest term in the denominator. This leads to the simple approximation

$$\frac{1 - F}{F} = \frac{[H^+]}{K_{a1}} + \cdots. \tag{3.43}$$

---

* This seems complicated, but if you examine Eq. (3.6), you will see that a sufficiently small value of A compared to C_T will cause V to be negative; this corresponds to titration with base of the same concentration as C_a. To prove this, derive the equation corresponding to (3.6) but assume that the titrant is strong base instead of strong acid. You will find an equation of the same form, but opposite sign.

The chemical basis for this approximation is that $[CO_3^=]$ is negligible from pH 5 to 8.3. Hence $C_T F = [HCO_3^-]$, $C_T(1 - F) = [CO_2]$, and the first equilibrium $[H^+][HCO_3^-] = K_{a1}[CO_2]$ yields the approximate expression (3.43). This result can also be obtained from (3.9) by neglecting the term in K_{a2} compared with the term in K_{a1}.

Substituting Eq. (3.43) in (3.41) gives

$$f_2 = (V_2 - V)[H^+] = (V - V_1)K_{a1} + \cdots, \tag{3.44}$$

and the equivalence point $V = V_1$ is found by extrapolating to $f_2 = 0$. As before, if $[H^+]$ is replaced by 10^{-pH}, the slope becomes $K_{a1}\gamma_+$ but is normally constant through this portion of the titration, since $[HCO_3^-]$ is replaced by $[Cl^-]$ in the expression for ionic strength.

This linearized plot is illustrated in part (b) of Fig. 3.4. The disadvantage of this plot is that V_2 must be known (presumably from a plot like the one in part (a) of Fig. 3.4), and whatever errors are incorporated in V_2 will be propagated in finding V_1. It is, of course, essential that 10 to 20 uniformly spaced pH values be taken over the entire range from 8.3 to about 3.5 in order for this analysis to be accurate.

For comparison, $[H^+]$ is also plotted in part (b) of Fig. 3.4. It shows pronounced curvature, and extrapolation would be unlikely to give an accurate value of V_1.

OTHER LINEARIZED PORTIONS OF THE ALKALINITY TITRATION CURVE

Once V_1 and V_2 are known, at least approximately, other transformations of the titration curve become possible, and these extrapolations may be used to verify the end points. Such verification is especially important in the presence of small concentrations of weak acids (ammonium ion, ferric ion, aluminum, borate).

To obtain V_2 from the high pH end, Eq. (3.44) is used in a rearranged form:

$$f_3 = \frac{V - V_1}{[H^+]} = \frac{V_2 - V}{K_{a1}} + \cdots. \tag{3.45}$$

This plot is linear (Fig. 3.5) with slope $1/K_{a1}$ and intercept $V = V_2$ at $f_3 = 0$. As before, the accuracy of extrapolation will depend on the accuracy of V_1.

The analogous function for determining V_1 by extrapolation from the alkaline side

$$f_4 = \frac{V - 2V_1 + V_2}{[H^+]} = \frac{V_1 - V}{K_{a2}} + \cdots \tag{3.46}$$

is plotted in Fig. 3.5. You will notice that it does not yield a good linear extrapolation because in its derivation G was neglected in Eq. (3.6). (Fortuitously, 10^{+pH} alone gives a better straight line!)

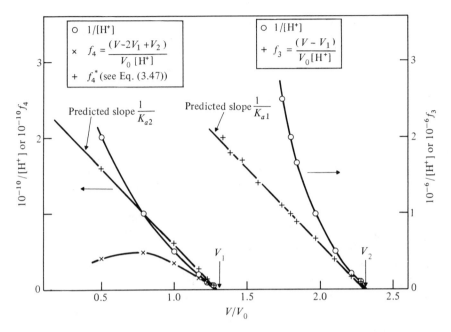

Fig. 3.5.   Some less frequently used Gran plots: f_3 extrapolates to the CO_2 equivalence point from the pH range 4.5–8.3; f_4^* extrapolates to the HCO_3^- equivalence point from the pH range above 8.3.

Use of Eq. (3.6) with appropriate approximations ($G = -K_w/[H^+]$; neglect $[H^+]^2$ in the denominator of F) yields a better expression:

$$f_4^* = f_4 + \frac{V + V_0}{C} \cdot \frac{K_w}{[H^+]^2} \cdot \frac{[H^+] + K_{a2}}{K_{a2}} = \frac{V_1 - V}{K_{a2}}. \tag{3.47}$$

As you can see from the plot in Fig. 3.5, this function does indeed yield a straight line, and you can also see from the plot that f_4^* is to a large extent governed by the term neglected in deriving Eq. (3.46).

THE $CO_3^=$ EQUIVALENCE POINT

Referring to Fig. 3.1, you will see that for a 10^{-3} M solution, there is no visible inflection of the titration curve at the equivalence point corresponding to Na_2CO_3, and hence no practical way to determine the end point. Fortunately, the other two end points give all the information needed to obtain A and C_T, so that the carbonate end point is not necessary. In sufficiently concentrated solutions, however, there is an inflection, and with excess base, a linear plot of $g = (V + V_0)10^{+pH}$ versus V can be obtained. It extrapolates to give a carbonate end point at $g = 0$.

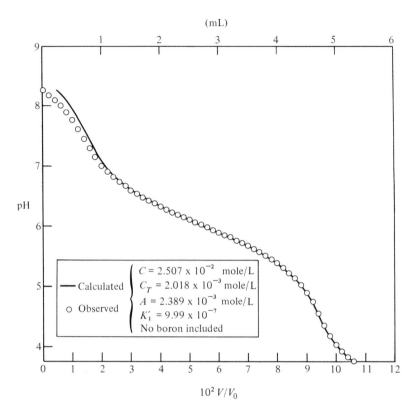

Fig. 3.6. Alkalinity titration of seawater. Circles are experimental data (Table 3.1) and the curve was calculated from Eqs. (3.6), (3.11), and (3.12). The term for borate (see Eqs. (3.51) to (3.53)) was not included for this plot. Parameters were: $C = 2.507 \times 10^{-2}$, $C_T = 2.018 \times 10^{-3}$, $A = 2.389 \times 10^{-3}$, $pK_1' = 6.000$, $pK_2' = 9.115$ (see Table 5.5).

ALKALINITY TITRATION IN SEAWATER

Now that you have seen how the titration curve and alkalinity are obtained for a simple model solution (CO_2 + NaOH), consider a natural water. Seawater is a more complicated mixture, and its response to titration with strong acid shows important differences from a carbonate–bicarbonate solution.

Figure 3.6 shows data from a titration of seawater with 0.02507 M hydrochloric acid.* Since the data are much more precise than can be shown on a graph, they are given to their full precision in Table 3.1.

---

* The seawater was collected near Bermuda on July 16, 1979, and the titration was performed the same day by Dr. Roland Wollast from the University of Brussels, who was working at the Bermuda Biological Station at the time. The salinity was not measured directly, but surface water from the Sargasso sea normally has salinity less than 36.5‰.

Table 3.1
Titration of seawater with standard hydrochloric acid

HCl concentration $C_a = 0.02507$ M, $T = 25.0°C$, $V_0 = 50.00$ mL)

V (mL)	pH	V (mL)	pH	V (mL)	pH
0	8.256	2.1	6.280	4.2	5.236
0.1	8.177	2.2	6.235	4.3	5.144
0.2	8.099	2.3	6.193	4.4	5.032
0.3	8.005	2.4	6.151	4.5	4.908
0.4	7.896	2.5	6.109	4.6	4.750
0.5	7.766	2.6	6.067	4.7	4.563
0.6	7.616	2.7	6.026	4.8	4.363
0.7	7.451	2.8	5.985	4.9	4.187
0.8	7.289	2.9	5.944	5.0	4.040
0.9	7.145	3.0	5.903	5.1	3.925
1.0	7.018	3.1	5.859	5.2	3.834
1.1	6.910	3.2	5.816	5.3	3.754
1.2	6.817	3.3	5.772	5.4	3.684
1.3	6.735	3.4	5.726	5.5	3.626
1.4	6.661	3.5	5.678	5.6	3.575
1.5	6.597	3.6	5.626	5.7	3.526
1.6	6.536	3.7	5.572	5.8	3.485
1.7	6.479	3.8	5.516	5.9	3.446
1.8	6.426	3.9	5.456	6.0	3.409
1.9	6.375	4.0	5.390	6.1	3.377
2.0	6.326	4.1	5.318	6.2	3.347

A Gran plot of the last 18 points (where strong acid is in excess) using f_1 (see Eq. (3.39))

$$f_1 = \frac{V + V_0}{V_0} 10^{-pH} = \gamma_+ \left(C_a \frac{V}{V_0} - A_T \right) + \cdots \tag{3.48}$$

is shown in Fig. 3.7, and the extrapolation to an intercept at $V_2 = 4.764$ is indicated. This corresponds to a total alkalinity of

$$A_T = \frac{V_2}{V_0} C_a = 2.385 \cdot 10^{-3} \text{ mole/L.}$$

The slope of this plot (according to (3.48)) gives

$$\gamma_+ = \frac{df_1}{dV} \frac{V_0}{C_a} = 0.690,$$

which agrees well with the Davies equation (2.36) for ionic strength $0.7 (35‰)$ salinity).

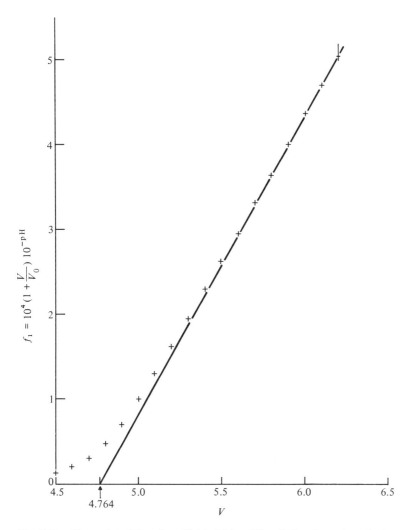

Fig. 3.7. Gran plot of data from Table 3.1 for pH < 5, showing extrapolation to $V_2 = 4.764$. This corresponds to $A = 2.389 \times 10^{-3}$ mole/L or 2.333×10^{-3} mole/kg.

Most of the first part of the curve can be linearized by the Gran function f_2 (see Eq. (3.44)):

$$f_2 = \frac{V_2 - V}{V_0} 10^{-\mathrm{pH}} = K'_1\left(\frac{V}{V_0} - \frac{V_1}{V_0}\right) + \cdots , \tag{3.49}$$

which is plotted in Fig. 3.8. The best linear fit to the data gives an intercept a at $V_1 = 0.630$ and a slope of $V_0 df_2/dV = 0.927 \cdot 10^{-6}$. According to (3.49), this slope

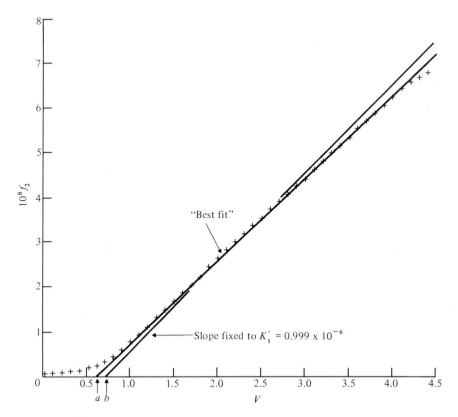

Fig. 3.8.   Gran plot of data from Table 3.1 for pH > 5 showing three choices of slope and intercept. The slope of the *best fit* line (corresponding to $K'_1 = 0.927 \cdot 10^{-8}$) does not agree with literature data for the equilibrium constant, but forcing the slope to the literature value $(0.999 \cdot 10^{-6})$ gives a poor fit.

should be K'_1, but the best literature value (Table 5.5) is 7% higher: $K'_1 = 0.999 \cdot 10^{-6}$. If you fix the slope at this latter value and make the line go through points near the middle of the graph, you will obtain an intercept b at $V_1 = 0.740$.

The total carbonate is obtained from the difference in endpoints (Eqs. (3.13) and (3.40)):

$$C_T = C_a \left(\frac{V_2}{V_0} - \frac{V_1}{V_0} \right) = 2.01 \cdot 10^{-3} \text{ mole/L.} \tag{3.50}$$

(Note that the "best-fit" line gives $C_T = 2.07 \cdot 10^{-3}$.)

With these parameters and the literature values of $K'_1 = 10^{-6.000}$, $K'_2 = 10^{-9.115}$ (Table 5.5) for seawater at 25.0°C, 35‰ salinity, the titration curve can be calculated from Eqs. (3.6) and (3.11). The experimental data and calculated curve

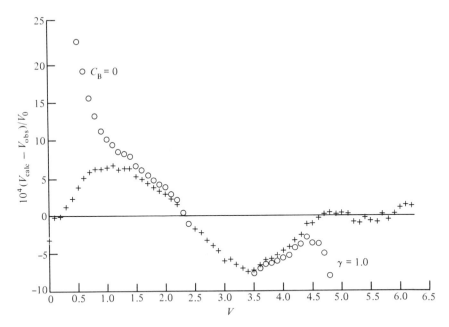

Fig. 3.9. Calculated versus observed values for the titration curve of Fig. 3.6. Parameters were as in Fig. 3.6 for the circles; for the crosses, the borate correction (Eq. (3.51)) was made by using $C_B = 0.476 \times 10^{-3}$, $pK'_B = 8.71$, and γ_+ was set at 0.691 instead of 1.0. These two additions correct the major deviations, but still leave residual systematic deviations 10 times as large as the precision of the measurements.

are directly compared in Fig. 3.6; the difference between the calculated and observed values of V/V_0 is shown on a magnified scale in Fig. 3.9.

At the beginning of the titration (Fig. 3.6), there are large deviations of the calculated from the observed values. These result from the failure of Eqs. (3.6) and (3.11) to take account of the significant amounts of borate in seawater. Borate can accept protons:

$$K'_B = \frac{[B(OH)_4^-]10^{-pH}}{[B(OH)_3]} = 10^{-8.71}. \tag{3.51}$$

The K'_B value (Table 5.5) is for 25°C and 19‰ chlorinity (approximately the same composition as 35‰ salinity). Thus at pH = 8.2, there is a substantial fraction of ionized borate. By Eq. (3.34), we have $A_T = A_c + [B(OH)_4^-]$. The alkalinity titration equation must therefore be modified by the inclusion of an extra term corresponding to $[B(OH)_4^-]$:

$$\frac{V}{V_0} = \frac{A - C_T F - C_B F_B + G'}{C - G'}, \tag{3.52}$$

where

$$F_B = \frac{K'_B}{K'_B + 10^{-pH}} \tag{3.53}$$

and $C_B = [B(OH)_4^-] + [B(OH)_3]$ is the total concentration of borate in solution.

An independent analysis of seawater for borate* gave $0.476 \cdot 10^{-3}$ mole/L. With the borate term included, the initial points of the calculated titration curve fell into excellent agreement with the observed values (Fig. 3.9).

Even with these important corrections, however, the deviation of the calculated from observed values over most of the titration curve far exceeds the precision of the measurements, and a clear systematic trend is apparent in the crosses plotted on Fig. 3.9: positive deviations in the first part of the titration curve, good agreement in the middle, and negative deviations until the equivalence point is reached.

Can these deviations be explained by other components of seawater? Silica cannot be the cause, since by analogy with boron (Eqs. (3.52) and (3.53)), $pK'_{Si} = 9.4$ predicts

$$F_{Si} = \frac{10^{-9.4}}{10^{-9.4} + 10^{-8.26}} = 0.07$$

at the beginning of the titration. C_{Si} is of the order of 2 to $5 \cdot 10^{-6}$ mole/L,[†] so that the effect of $C_{Si}F_{Si}$ on V/V_0 is not more than about $4 \cdot 10^{-7}$, and then only at the beginning of the titration. This effect is totally buried by the effect of borate.

Phosphate has acidity constants in the right range: $pK'_2 = 6.08$ and $pK'_3 = 8.58$ in seawater at 25°C. An additional term $-C_P F_P$ would be introduced in the numerator of (3.16), where

$$F_P = \frac{[HPO_4^=] + 2[PO_4^\equiv]}{C_P} = \frac{10^{-pK'_2 - pH} + 10^{-pK'_2 - pK'_3 + 0.3}}{10^{-2pH} + 10^{-pK'_2 - pH} + 10^{-pK'_2 - pK'_3}}. \tag{3.54}$$

This function varies smoothly from about 1.3 at the beginning of the titration to less than 0.01 at the equivalence point, so that the term $F_P C_P$ cannot affect V/V_0 by more than $F_P C_P/(C - G')$ or approximately $50 C_P$. Typical values for C_P of seawater in the open ocean off Bermuda are 1 to $2 \cdot 10^{-7}$ mole/L, and even the Bermuda Biological Station seawater supply in the labs does not exceed $2 \cdot 10^{-6}$ mole/L phosphorus.[†] Thus the maximum effect on V/V_0 is less than 10^{-4}, and more typically of the order of 10^{-5}. The systematic deviations shown in Fig. 3.9 are much larger ($\pm 7 \cdot 10^{-4}$) and have not been adequately explained.[‡]

---

* Skirrow, op. cit., p. 29, recommends $C_B = 2.05 \cdot 10^{-5}$ (Cl‰), lower than this estimate.

† T. Jickells and B. Von Bodungen (Bermuda Biological Station), personal communication. See also Menzel, D. and Ryther, J. 1961. Annual variation in primary production of the Sargasso Sea off Bermuda. *Deep-Sea Research* 7:282–288.

‡ Note that K'_1 increases slightly with salinity to $1.016 \cdot 10^{-6}$ at 36.5‰. In order to obtain $K'_1 = 0.927 \cdot 10^{-6}$, sufficient fresh water would have to be added to lower the salinity to 28.8‰. This seems unlikely.

THE BUFFER INDEX

This concept was first developed for mixtures of approximately equal concentrations of conjugate acid and base (i.e., 0.05 M $NaHCO_3$ and 0.05 M Na_2CO_3). Such buffer solutions strongly resist changes in pH when acids or bases are added, and are useful as pH standards. The buffer index tells how great this resistance to pH change will be.

In terms of the titration curve, the greater the pH change is for a given addition of acid or base, the lower is the buffer index. In Fig. 3.1, for example, the steepest parts (lowest buffer index) are at the two equivalence points and the least steep part (highest buffer index) is halfway between.

There are a number of possible ways to define buffer indexes mathematically. In this chapter we will consider two indexes*: the change in alkalinity with pH at constant total carbonate

$$\beta_C = \left(\frac{\partial A}{\partial \mathrm{pH}}\right)_{C_T} \tag{3.55}$$

and the change in alkalinity with pH at constant CO_2 partial pressure

$$\beta_P = \left(\frac{\partial A}{\partial \mathrm{pH}}\right)_{P_{CO_2}}. \tag{3.56}$$

A larger value for β means that more acid or base must be added to achieve the same change in pH. In Chapter 4, I will show how you take account of the effects of insoluble carbonate salts such as $CaCO_3$, and in Chapter 5 I will derive a *homogeneous buffer factor* that describes how changes in atmospheric P_{CO_2} affect the dissolved carbonate in the oceans at constant alkalinity.

The relation between $d\mathrm{pH}$ and $d[H^+]$ is obtained as follows:

$$\mathrm{pH} = -\log_{10}[H^+] - \log_{10}\gamma_+ = -\frac{\ln[H^+]}{2.303} - \frac{\ln\gamma_+}{2.303},$$

$$d\mathrm{pH} = -\frac{d[H^+]}{2.303[H^+]} - \frac{d\ln\gamma_+}{2.303} + \cdots.$$

Note that at constant ionic strength, γ_+ is constant, but even if I varies with pH, the derivative $d\ln\gamma_+/d\mathrm{pH}$ is normally negligible (see Problem 24). Therefore

$$d[H^+] = -2.303[H^+]d\mathrm{pH} + \cdots. \tag{3.57}$$

Since the mass balance–equilibrium combinations are expressed in terms of $[H^+]$, derivatives of these with respect to $[H^+]$ can be transformed to derivatives with respect to pH by using Eq. (3.57).

---

* These quantities β have the dimensions of concentration and are called *buffer intensity* and *buffer capacity* as well as *buffer index*.

BUFFER INDEX AT CONSTANT C_T

The first example is β_C, which is the inverse of the slope of the alkalinity titration curve, and hence closely related to the functions discussed earlier in this chapter. First recall the expression for alkalinity (Eq. (2.28)):

$$A = C_T \frac{K_{a1}[H^+] + 2K_{a1}K_{a2}}{K_{a1}K_{a2} + K_{a1}[H^+] + [H^+]^2} + \frac{K_w}{[H^+]} - [H^+]. \tag{3.58}$$

Take the derivative* of each term with respect to $[H^+]$, holding C_T and the equilibrium constants constant, and obtain

$$\left(\frac{\partial A}{\partial [H^+]}\right)_{C_T} = -C_T K_{a1} \frac{K_{a1}K_{a2} + 4K_{a2}[H^+] + [H^+]^2}{(K_{a1}K_{a2} + K_{a1}[H^+] + [H^+]^2)^2} - \frac{K_w}{[H^+]^2} - 1. \tag{3.59}$$

This expression is general for any diprotic acid, but because of the numerical values of K_{a1} and K_{a2}, the middle term of the numerator $(4K_{a2}[H^+])$ is less than 1% of the other terms for any pH from 0 to 14 (see Eq. (3.10) and Problem 25).

Alternatively, the form of F based on this approximation (Eq. (3.9)) can be used to obtain the derivative

$$\left(\frac{\partial A}{\partial [H^+]}\right)_{C_T} = -C_T\left(\frac{K_{a1}}{(K_{a1} + [H^+])^2} + \frac{K_{a2}}{(K_{a2} + [H^+])^2}\right) - \frac{K_w}{[H^+]^2} - 1 + \cdots. \tag{3.60}$$

Make use of Eq. (3.57), then (3.59), to get

$$\beta_C = \left(\frac{\partial A}{\partial pH}\right)_{C_T} = -2.303[H^+]\left(\frac{\partial A}{\partial [H^+]}\right)_{C_T}$$

$$\beta_C = 2.303 C_T K_{a1}[H^+] \frac{K_{a1}K_{a2} + [H^+]^2}{(K_{a1}K_{a2} + K_{a1}[H^+] + [H^+]^2)^2}$$

$$+ 2.303\left(\frac{K_w}{[H^+]} + [H^+]\right). \tag{3.61}$$

The alternative form derived from (3.60) can also be used. An illuminating modification is to use the combinations of equilibrium constants to reconstruct concentrations of chemical species (refer to Eqs. (2.21), (2.22), and (2.23)):

$$\beta_C = 2.303[HCO_3^-] \frac{[CO_3^=] + [CO_2]}{C_T} + 2.303([OH^-] + [H^+]). \tag{3.62}$$

---

* Refer to any elementary calculus book: if U and V are functions of X, then

$$\frac{d}{dX}\left(\frac{U}{V}\right) = \frac{1}{V^2}\left(V\frac{dU}{dX} - U\frac{dV}{dX}\right).$$

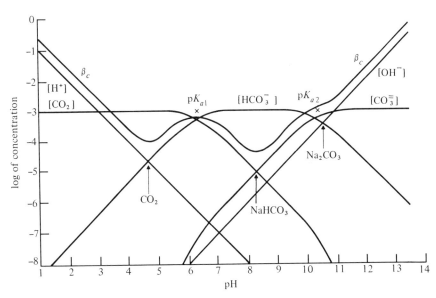

Fig. 3.10. Buffer index β_c for $C_T = 10^{-3}$ as a function of pH, superimposed on the logarithmic concentration diagram, to illustrate the contributions from the various terms in Eq. (3.62).

Note that this can also be derived from (3.60) by making suitable approximations (see Problem 26). For computational purposes, (3.60) or (3.61) are most useful; for graphical and conceptual purposes, the form of (3.62) is much more elegant.

In Fig. 3.10, the curve of β_C is plotted along with the logarithmic concentration diagram (refer back to Fig. 2.3), giving the concentrations of the various species at $C_T = 10^{-3}$. Note that at sufficiently low pH, β_C is determined by $[H^+]$, and at sufficiently high pH, by $[OH^-]$. The minima in β_C correspond to the CO_2 (pH $\cong$ 4.5) and HCO_3^- (pH $\cong$ 8.3) equivalence points of the alkalinity titration. The maximum at pH = 6.3 occurs when pH = pK_{a1} or $[CO_2] = [HCO_3^-]$. Note also that there is no minimum in β_C at pH = 10.5, corresponding to Na_2CO_3; this is in agreement with Fig. 3.1, where no inflection occurs in the titration curve.

BUFFER INDEX AT CONSTANT P

Making use of Eqs. (2.1), (2.2), and (2.3), you can obtain an expression for alkalinity as a function of pH at constant partial pressure of CO_2:

$$A = [CO_2] + 2[CO_3^=] + [OH^-] - [H^+], \tag{3.1}$$

$$A = \frac{K_{a1}K_H P_{CO_2}}{[H^+]} + \frac{2K_{a1}K_{a2}K_H P_{CO_2}}{[H^+]^2} + \frac{K_w}{[H^+]} - [H^+], \tag{3.63}$$

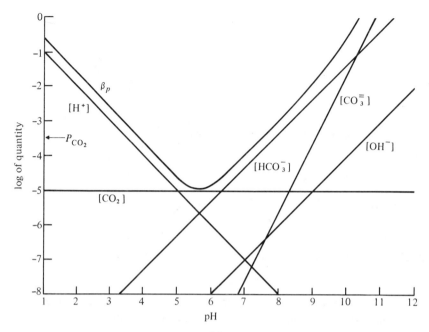

Fig. 3.11. Buffer index β_p for $P = 10^{-3.5}$, superimposed on a logarithmic concentration diagram, to illustrate the contributions of the various terms in Eq. (3.66).

and the derivative of this function gives

$$\beta_P = \left(\frac{\partial A}{\partial \text{pH}}\right)_{P_{CO_2}} = -2.303[\text{H}^+]\left(\frac{\partial A}{\partial [\text{H}^+]}\right)_{P_{CO_2}}. \tag{3.64}$$

The computations are similar to, but simpler than, the above for constant C_T and yield the following two alternative forms:

$$\beta_P = 2.303\left(\frac{K_{a1}K_H P_{CO_2}}{[\text{H}^+]}\right)\left(1 + \frac{4K_{a2}}{[\text{H}^+]}\right) + 2.303\left(\frac{K_w}{[\text{H}^+]} + [\text{H}^+]\right), \tag{3.65}$$

$$\beta_P = 2.303([\text{HCO}_3^-] + 4[\text{CO}_3^=] + [\text{OH}^-] + [\text{H}^+]). \tag{3.66}$$

This function is plotted in Fig. 3.11, with $P_{CO_2} = 10^{-3.5}$ atm. (This value makes $C_T = 10^{-5}$ at low pH.) Note that for pH less than 5.5, β_p and β_c are essentially identical in form, being composed of the same terms in $[\text{HCO}_3^-]$ and $[\text{H}^+]$. (In Eq. (3.62), $[\text{CO}_2] \cong C_T$ in this pH range.)

Over the pH range from 5 to 8, while β_c is going through a minimum, β_P increases linearly: this difference is caused by the uptake of CO_2 by the solution, which increases C_T as pH increases.

In the high pH range, β_P is controlled by $[\text{CO}_3^=]$; above pH = 10.5, NaHCO$_3$ crystallizes and Eqs. (3.65) or (3.66) do not apply. Note that at pH = 10 in a solution saturated with CO_2 at $P_{CO_2} = 10^{-3.5}$ atm, the buffer index is as high as it is at pH = 13 in a solution with constant C_T (Fig. 3.10).

PROBLEMS

1. Calculate the equivalence point pH for the titration of a water with alkalinity $2 \cdot 10^{-3}$ equivalents/L and pH = 7.5 with 0.1 M HCl.

   Answer: 4.5

2. Derive Eqs. (3.6), (3.7), and (3.8) from Eqs. (3.1) to (3.5) and Eqs. (2.21) and (2.22).

3. Calculate a curve of pH versus V for a water with $A = 2.3 \cdot 10^{-3}$, $C_T = 1.0 \cdot 10^{-3}$, titrated by 10^{-1} M NaOH. Compare with Fig. 3.1.

4. Plot $(1 - F)/F$ versus pH on an expanded scale for the pH range 7 to 9. Compare with Eq. (3.43).

5. Make suitable approximations of Eqs. (3.6), (3.7), and (3.8) for the pH range 7 to 9.

6. A solution of alkalinity $2.5 \cdot 10^{-3}$ and total carbonate $2.0 \cdot 10^{-3}$ M is titrated with 0.0100 M HCl solution. The only CO_2 in the atmosphere above the solution comes from the solution itself (e.g., because the air above the beaker was purged with nitrogen and sealed with Saran Wrap):

   a) Calculate the pH at the start of the titration.

   b) Find pH and partial pressure of CO_2 at the equivalence point.

   c) Estimate the titration error (% of V/V_0) if the endpoint is taken to be pH = 4.60.

   Answer: (a) pH = 9.85, (b) pH = 4.52, $P_{CO_2} = 0.059$ atm, (c) -0.17%.

7. If a solution with $A = 5 \cdot 10^{-4}$ was titrated with acid, $C_a = 10^{-2}$, and the temperature was 5°C ($pK_{a1} = 6.52$), but the pH at the endpoint was taken to be 4.50, what would be the titration error?

   Answer: If $C_T \le A$, $E = +5.4\%$; otherwise see Eq. 3.19.

8. An industrial waste from a metals industry contains approximately $5 \cdot 10^{-3}$ M H_2SO_4. Before being discharged into the stream, the water is diluted with tap water in order to raise the pH. The tap water has the following composition: pH = 6.5, alkalinity = $2 \cdot 10^{-3}$ equivalents/L, $[Ca^{++}] = 2 \cdot 10^{-3}$ M, $[Mg^{++}] = 10^{-4}$ M, $[Cl^-] = 10^{-3}$ M. What dilution is necessary to raise the pH to approximately 4.5 ± 0.4? (Note that H_2SO_4 is a strong acid, i.e., K_1 and K_2 are larger than 10^{-2}.)

   Answer: 5 times dilution

9. 50.0 mL seawater was titrated with 0.02507 M HCl at 25.0°C. Some of the data are

V (mL)	pH	V (mL)	pH	V (mL)	pH
4.0	5.390	5.0	4.040	5.8	3.485
4.5	4.908	5.2	3.834	6.0	3.409
4.6	4.750	5.4	3.684	6.2	3.347
4.7	4.563	5.6	3.575		

Find the equivalence point and hence the alkalinity by using the Gran plot of Eq. (3.39). Knowing that the initial pH was 8.256 and the total borate concentration was $4.76 \cdot 10^{-4}$ mole/L, calculate the carbonate alkalinity.

Answer: $A_T = 2.39 \cdot 10^{-3}$ mole/L, $A_c = 2.26 \cdot 10^{-3}$ mole/L.

10. Some more data from the titration of Problem 9 are

V (mL)	pH	V (mL)	pH
0.0	8.256	1.4	6.661
0.5	7.766	1.6	6.536
1.0	7.018	1.8	6.426
1.2	6.817	2.0	6.326

Use the Gran plot of Eq. (3.44) together with the answer to Problem 9 to obtain C_T.

Answer: $2.02 \cdot 10^{-3}$ mole/L.

11. A solution contains 10^{-3} M $NaHCO_3$ and 10^{-3} M Na_2CO_3. Calculate the pH of the solution and its buffer capacity β_C.

Answer: pH $= pK'_{a2} = 10.36$, $\beta_C = 1.15 \cdot 10^{-3}$

12. Find the buffer index β_P of a solution containing CO_2 at partial pressure 10^{-2} atm in water of alkalinity $3 \cdot 10^{-3}$ equivalents/L.

Answer: $6.9 \cdot 10^{-3}$

13. A source of error in acid–base titrations is the possible presence of carbonate. If a 10^{-3} M HCl solution were titrated with 10^{-3} M NaOH, the equivalence point would be expected at pH $= 7.00$. If this HCl had absorbed CO_2 from the atmosphere and contained $[CO_2] = 10^{-5}$ M, find the titration error if the end point were taken at pH $= 7.0$. Compare with the titration error if the end point were taken at pH $= 5.0$.

Answer: 1%; less than 0.1%

14. Alkalinity can be measured (Gripenberg, 1937. 5th Hydrol. Conf. Helsingfors, Comm. 10B.) by treating a sample with a measured excess of strong acid, driving off the liberated CO_2 by purging with N_2 or boiling, and back-titrating with strong base to pH of 6.0 to 7.0 (where, for example, borate is negligibly ionized). Show that this method gives the correct result. (See Skirrow, G., 1975. *Chemical Oceanography*, vol. 2, Riley, J. P. and Skirrow, G., eds. New York: Academic Press, p. 27.)

15. Culberson's (Culberson, C., Pytkowicz, R. M. and Hawley, J. E., 1970. *J. Mar. Res.* 28:15) modification of the Gripenberg method consists of adding 30 mL of 0.01 M HCl to a 100 mL sample of sea water. The CO_2 is purged with CO_2-free, water-saturated air, and the final pH is measured. Show that the alkalinity is given by

$$A = \frac{1000}{V_s} V M - \frac{1000}{V} (V_s + V) \frac{10^{-pH}}{\gamma_H},$$

where V_s is the sample volume, V is the HCl volume, M is the molar concentration of HCl, and γ_H is the activity coefficient of hydrogen ion (see Skirrow, op. cit., p. 28).

16. Acid rain containing C_T mole/L of CO_2, C_S mole/L of H_2SO_4, C_A mole/L of HNO_3, and C_N mole/L of NH_4^+ is titrated with strong base of concentration C_b. Derive an equation analogous to (3.6) for the titration curve. Plot a sample titration curve for $C_T = 10^{-5}$, $C_S = 2 \cdot 10^{-5}$, $C_A = 10^{-5}$, $C_N = 10^{-5}$. (For H_2SO_4, $pK_{a1} < 0$, $pK_{a2} = 1.8$; for HNO_3, $pK_a < 0$; for NH_4^+, $pK_a = 9.25$.)

17. If a Gran plot (Eq. (3.44)) were used to determine the HCO_3^- end point of the titration described in Problem 16, estimate the error in V_1/V_0 caused by the presence of sulfate, nitrate, and ammonium ion at the concentrations given.

18. Show that the relationship

$$F = \frac{[HCO_3^-] + 2[CO_3^=]}{C_T} = \frac{K_1' 10^{-pH} + 2K_1' K_2'}{K_1' K_2' + K_1' 10^{-pH} + 10^{-2pH}}$$

holds when hybrid constants are used at constant ionic strength.

19. Using Eq. (3.22), calculate pH at the equivalence point of the alkalinity titration in the presence of 0.5 M NaCl. If the end point of a titration with $C_T = A = 10^{-3}$ at 25°C were taken at a pH calculated with the equilibrium constants for zero ionic strength, how much error would result?

   Answer: pH $= 4.651$; $E = 0.37\%$

20. Show that the end point determined by the Gran titration for alkalinity in the presence of a noncomplexing ionic medium is the same as the equivalence point.

21. A natural water with $A = 10^{-4.5}$ and $C_T = 10^{-4}$ is titrated with 0.01 M HCl until $[Cl^-] = 10^{-3}$. Calculate $f_1 = (V + V_0)10^{-pH}$ and $f_1' = (V + V_0)[H^+]$ for five to ten points at pH < 5. Use the Davies equation (2.36) to relate pH and $[H^+]$. Plot f_1 and f_1' versus V and extrapolate by using a straightedge to $f_1 = 0$ and $f_1' = 0$. Estimate the maximum error that would be incurred by using f_1 instead of f_1'.

22. Protonation of sulfate ($K_a' = 10^{-1.1} = 10^{-pH}[SO_4^=]/[HSO_4^-]$) could modify the shape of the Gran plot in seawater (such as Fig. 3.7). What error in the intercept would result from forcing a straight line to fit the data? How could it be minimized?

23. Derive equation (3.52) and the related equations that give the form of the alkalinity titration curve in the presence of borate, silicate, and phosphate (compare Eq. (3.54)).

24. Estimate the error in Eq. (3.57) resulting from neglecting the derivative of the activity coefficient γ_+. Assume that the pH of pure water is being changed from 5.1 to 5.0 by addition of HCl.

   Answer: $d[H^+] = -2.303[H^+](dpH + d\log \gamma_+)$. For the values given, $dpH = 0.10$ and $d\log \gamma_+ = 1.7 \cdot 10^{-4}$. The error is 0.17% of $d[H^+]$.

25. In deriving Eq. (3.61), the term $4K_{a2}[H^+]$ was neglected compared to $K_{a1}K_{a2}$ and $[H^+]^2$. Show that this introduces an error of no more than 1% over the pH range from 0 to 14.

26. Make suitable approximations on the expressions for $[CO_2]$, $[HCO_3^-]$, and $[CO_3^=]$ at constant C_T, so as to derive Eq. (3.62) from Eq. (3.60). Estimate the magnitude of error incurred in β_C by these approximations.

27. The expression for total alkalinity of seawater (see Chapter 5) contains additional terms compared to Eq. (2.16) or (3.1). The largest of these terms is due to borate (see Eq. (3.52)):

$$A_T = [HCO_3^-] + 2[CO_3^=] + [OH^-] - [H^+] + [H_2BO_3^-],$$

where total borate is

$$C_B = [H_3BO_3] + [H_2BO_3^-].$$

The acidity constant of boric acid is expressed as (see Eq. (3.51)):

$$K_B' = \frac{[H_2BO_3^-] \cdot 10^{-pH}}{[H_3BO_3]} = 10^{-8.71}.$$

The acidity constants for CO_2 are expressed as (compare Eqs. (2.39) and (2.42)):

$$K_1' = \frac{[HCO_3^-] \cdot 10^{-pH}}{[CO_2]}, \qquad K_2' = \frac{[CO_3^=] \cdot 10^{-pH}}{[HCO_3^-]}.$$

Show that

$$A_T = x + 2y + C_B\left(1 + \frac{\beta_2 x}{y}\right)^{-1}, \qquad C_T = x + y + \frac{\beta_1 x^2}{y},$$

where $x = [HCO_3^-]$, $y = [CO_3^=]$, $\beta_1 = K_2'/K_1'$, $\beta_2 = K_2'/K_B'$ (see Keir, R. S. 1979, *Marine Chemistry*. 8:95–97). Elimination of x without approximations leads to a cubic in y:

$$py^3 + qy^2 + ry + s = 0.$$

Evaluate the coefficients p, q, r, and s, both algebraically and numerically, with pH = 8, $A = C_T = 2 \cdot 10^{-3}$, $C_B = 4 \cdot 10^{-4}$. Develop an iterative strategy to obtain y. For example, compare the use of Newton's methods with a quadratic solution assuming $p = 0$ initially.

28. In Fig. 3.8, you saw the use of the Gran function f_2 to obtain the HCO_3^- end point V_1. Recall from the earlier discussion (p. 57) that this end point can also be obtained by extrapolation of f_3 (Eq. (3.45)) in the solution that contains only CO_2 and strong base. In seawater, this second extrapolation can also be made. Test to see whether it gives a consistent value for V_1. Use the full equation for total alkalinity to account for any discrepancies.

CHAPTER FOUR

Solubility Equilibria: Calcium Carbonate

Many applications of carbonate equilibria involve insoluble carbonate salts. By far the most common is $CaCO_3$, which is the principal component of the earth's limestone sediments and rocks as well as the skeletons of most marine invertebrates. There are two common calcium carbonate minerals, calcite and aragonite. Under conditions of perfect equilibrium at ambient temperature and pressure, calcite is the more stable phase, but aragonite is often the phase deposited biologically, and the conversion to calcite occurs slowly. Dolomite, $CaMg(CO_3)_2$, and magnesian calcites with $Mg/Ca < 1$ are commonly found in rocks derived from marine environments (more details will be given in Chapter 5).

The fundamental equilibrium expression is the solubility product:*

$$[Ca^{++}][CO_3^=] = K_{s0}. \tag{4.1}$$

If the product of concentrations is less than K_{s0}, the solid will dissolve; if the product in solution exceeds K_{s0}, the solid will precipitate until the equality is satisfied. Values of K_{s0} at several ionic strengths are given in Table 2.1 and Fig. 2.9. Values of K_{s0}^0 at temperatures from 0 to 50°C are given in Table 2.2. Indeed, two sets of values are given: one set, the traditional values (that is, $pK_{s0}^0 = 8.34$ at 25°C), and the other set, values recalculated[†] with a model that took account of ion pairs, especially $CaHCO_3^+$.

As early as 1941, the $CaHCO_3^+$ ion pair was postulated and its association constant was measured.[‡] A good recent value (P. B. Hostetler, 1973, cited by Christ, et al., loc. cit.) is

$$[CaHCO_3^+]\gamma_+ = 10^{+1.22}[Ca^{++}][HCO_3^-]\gamma_{++}\gamma_=, \tag{4.2}$$

---

* As in Chapters 1–3, concentration equilibrium constants are used for simplicity. The activity product (sometimes denoted K_{sp}) is the value of K_{s0} extrapolated to zero ionic strength, so I use the notation $K_{s0}^0 = [Ca^{++}][CO_3^=]\gamma_{++}\gamma_=$ (see Eqs. 2.50 and 2.51).

† Christ, C. L., Hostetler, P. B., and Siebert, R. M. 1974. *J. Res. US Geol. Survey* 2:175–184.

‡ Greenwald, I. 1941. *J. Biol. Chem.* 141:789; Garrels, R. M. and Thompson, M. E. 1962. *Am. J. Sci.* 260:57–66. An argument for continuing to omit $CaHCO_3^+$ from chemical models was made by Plummer, L. N. and Mackenzie, F. T. 1974. *Am. J. Sci.* 274:61–83.

and this was used in Christ et al.'s (loc. cit.) recalculation of K_{s0}^0. Their values are significantly different from the traditional values: for example, $pK_{s0}^0 = 8.52 \pm 0.04$ instead of 8.34 at 25°C.

The easiest way to see the relative importance of this ion pair and of the other two ion pairs that might be significant ($CaOH^+$ and uncharged, dissolved $CaCO_3$) is in the form of a logarithmic concentration diagram, such as was used in Chapter 2. As before, the carbonate species are functions of $[H^+]$ and of the partial pressure of CO_2. For simplicity, 25°C zero ionic strength constants from Table 2.2 are used in the numerical calculations:

$$[CO_2] = K_H P = 10^{-1.47} P, \qquad \text{(2.1) or (4.3)}$$

$$[HCO_3^-] = \frac{K_{a1} K_H P}{[H^+]} = 10^{-7.82} \frac{P}{[H^+]}, \qquad \text{(2.5) or (4.4)}$$

$$[CO_3^=] = \frac{K_{a1} K_{a2} K_H P}{[H^+]^2} = 10^{-18.15} \frac{P}{[H^+]^2}. \qquad \text{(2.6) or (4.5)}$$

From (4.1), with $pK_{s0} = 10^{-8.52}$, together with (4.5), you can get

$$[Ca^{++}] = \frac{K_{s0}[H^+]^2}{K_{a1} K_{a2} K_H P} = 10^{+9.63} \frac{[H^+]^2}{P}. \qquad \text{(4.6)}$$

Then substitute (4.6) and (4.4) in (4.2) ($K = 10^{+1.22}$):

$$[CaHCO_3^+] = \frac{K K_{s0}[H^+]}{K_{a2}} = 10^{+3.03}[H^+]. \qquad \text{(4.7)}$$

The next ion pair is given by

$$[CaCO_3^0] = 10^{+3.1}[Ca^{++}][CO_3^=] = 10^{+3.1} 10^{-8.52} = 10^{-5.42}. \qquad \text{(4.8)}$$

($[CaCO_3^0] = K_{s1}$ is the product of the ion association constant and K_{s0}. It is independent of both P and $[H^+]$, provided solid $CaCO_3$ is in equilibrium with the solution.*) Finally,

$$[CaOH^+] = 10^{+1.3}[Ca^{++}][OH^-] = 10^{-3.07} \frac{[H^+]}{P}. \qquad \text{(4.9)}$$

All these equations are displayed in Fig. 4.1 for $P = 10^{-3.5}$ atm and in Fig. 4.2 for $P = 1.0$ atm. At the lower (average atmospheric) partial pressure, Ca^{++} is by far the most important calcium-containing species for pH < 9. As you already know

---

* The value quoted comes from Christ et al. (loc. cit.) and agrees with Garrels, R. M. and Christ, C. L. (1965. *Solutions, Minerals, and Equilibria.* New York: Harper & Row.). However, some workers believe K_{s1} is substantially larger; see Nakayama, F. 1969. *Soil. Sci. Soc. Am. Proc.* 33:668 for data leading to $K_{s1} \cong 10^{-4}$; see also Chapter 6, p. 215.

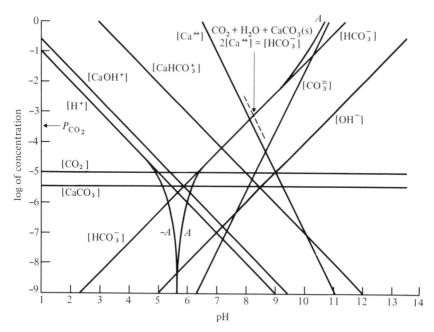

Fig. 4.1. Logarithmic concentration diagram for water saturated with $CaCO_3$ under a partial pressure $P = 10^{-3.5}$ atm of CO_2. Note that the intersection $2[Ca^{++}] = [HCO_3^-]$ corresponds to the condition where no acid or base is added other than CO_2 and $CaCO_3$. Note also that none of the ion pairs is significant compared to the major species under these conditions. However, under sufficiently basic conditions, the neutral ion pair $CaCO_3^0$ becomes the most important Ca species.

from Chapter 2, HCO_3^- is the most important carbonate-containing species from pH = 7 to 10. At the higher partial pressure, however, $[CaHCO_3^+]$ is 10% of $[Ca^{++}]$ at pH = 5.5, and this ion pair becomes the principal calcium-containing species between pH = 6.5 and 8.5. It is also the principal carbonate-containing species at pH < 4.5.

CALCIUM CARBONATE, CARBON DIOXIDE, AND WATER

Addition of carbon dioxide to water in equilibrium with solid calcium carbonate will cause the solid to dissolve. It will produce all the ions discussed above, but the principal ones will be Ca^{++} and HCO_3^-. The charge balance is

$$[H^+] + 2[Ca^{++}] + [CaOH^+] + [CaHCO_3^+]$$
$$= [HCO_3^-] + 2[CO_3^=] + [OH^-]. \tag{4.10}$$

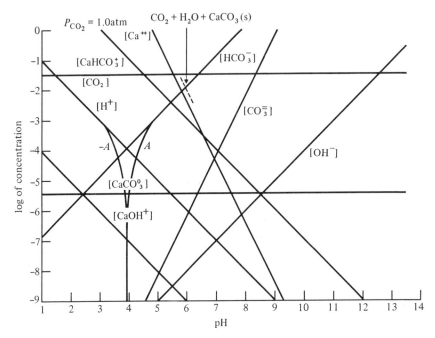

Fig. 4.2. Logarithmic concentration diagram for water saturated with $CaCO_3$ under a CO_2 pressure of 1 atm. The intersection $2[Ca^{++}] = [HCO_3^-]$ still determines the pH of the solution with no added acid or base, but the ion pair $CaHCO_3^+$ makes a substantial contribution (about 10%) to the mass and charge balances. At higher pH, $[CaHCO_3^+]$ exceeds $[Ca^{++}]$, and at the highest pH range, $CaCO_3^0$ is the most important calcium-containing dissolved species.

As pointed out above, $2[Ca^{++}]$ is normally the largest term on the left and $[HCO_3^-]$ is the largest term on the right:

$$2[Ca^{++}] = [HCO_3^-] + \cdots. \tag{4.11}$$

EXAMPLE 1   To find the pH (intersection marked on Fig. 4.1) numerically, substitute (4.4) and (4.6) in (4.11):

$$(10^{+0.3})(10^{+9.63})\frac{[H^+]^2}{P} = 10^{-7.82}\frac{P}{[H^+]} + \cdots,$$

$$[H^+] = 10^{-5.92}P^{2/3} + \cdots, \tag{4.12}$$

or pH = 8.25 at $P = 10^{-3.5}$ if $\gamma_+ \cong 1.0$.

   At $P = 1.0$ (Fig. 4.2) you will note that $[CaHCO_3^+]$ is about 10% of $2[Ca^{++}]$, where the intersection is marked. A more complete approximation to (4.10) is therefore

$$2[Ca^{++}] + [CaHCO_3^+] = [HCO_3^-] + \cdots. \tag{4.13}$$

Substitute (4.4), (4.6), and (4.7) in (4.13):

$$10^{+9.93}\frac{[H^+]^2}{P} + 10^{+3.03}[H^+] = 10^{-7.82}\frac{P}{[H^+]} + \cdots, \tag{4.14}$$

then rearrange (4.14) to resemble (4.12):

$$[H^+] = (10^{-17.75}P^2 - 10^{-6.90}[H^+]^2P)^{1/3} + \cdots. \tag{4.15}$$

This cubic can be solved by successive approximations. The first approximation is to neglect $[CaHCO_3^+]$; that is, use (4.12) with $P = 1.0$ and obtain $[H^+] = 10^{-5.92}$. Then substitute this value in the right-hand side of (4.15), and evaluate the expression to give the second approximation:

$$[H^+] = (10^{-17.75} - 10^{-6.90}(10^{-5.92})^2)^{1/3} = 10^{-5.93}.$$

Thus the inclusion of the ion pair has affected the pH by only 0.01 unit.   ∎

The alkalinity of this solution is obtained from the usual definition

$$A = [HCO_3^-] + 2[CO_3^=] + [OH^-] - [H^+]. \qquad \text{(2.16) or (4.16)}$$

At the partial pressures and pH range we have been considering, only $[HCO_3^-]$ is important in (4.16) (see Figs. 4.1 and 4.2), hence

$$A = [HCO_3^-] + \cdots. \tag{4.17}$$

Equation (4.4) gives $[HCO_3^-]$ in terms of P and $[H^+]$; Eq. (4.12) gives $[H^+]$ as a function of P, so that

$$A = \frac{10^{-7.82}P}{10^{-5.92}P^{2/3}} = 10^{-1.90}P^{1/3} + \cdots, \tag{4.18}$$

or $A = 10^{-1.90}$ at $P = 1$, $A = 10^{-3.07}$ at $P = 10^{-3.5}$.

EXAMPLE 2   What is the effect of ionic strength (pp. 30–40)? Use the full equilibrium expressions corresponding to (4.4) and (4.6) in (4.11) and obtain

$$\frac{2K_{s0}^0[H^+]^2\gamma_+^2}{K_{a1}^0K_{a2}^0K_H^0P\gamma_{++}} = \frac{K_{a1}^0K_H^0P}{[H^+]\gamma_+\gamma_-} + \cdots,$$

$$10^{-pH} = [H^+]\gamma_+ = \left(\frac{K_{a1}^{02}K_{a2}^0K_H^{02}\gamma_{++}}{2K_{s0}^0\gamma_-}P^2\right)^{1/3} + \cdots. \tag{4.19}$$

By (4.17) and (4.10), $A = [HCO_3^-]$ and $A = 2[Ca^{++}] + \cdots$,

$$I = \tfrac{1}{2}([HCO_3^-] + 4[Ca^{++}] + \cdots) = \tfrac{3}{2}A + \cdots. \tag{4.20}$$

From the Davies equation (2.36) you get $\gamma_+ = 10^{-0.059}$ at $P = 1$ and $\gamma_{++} = 10^{-0.017}$ at $P = 10^{-3.5}$.

Using the equilibrium constants for 25°C and $\gamma_{++}/\gamma_- = \gamma_+^3$, you get pH = 6.11 (instead of 5.93) at $P = 1.0$; pH = 8.30 (instead of 8.25)* at $P = 10^{-3.5}$. (These errors are comparable to those tolerated for years because pK_{s0}^0 was taken to be 8.34 instead of 8.52, which causes pH to be underestimated by about 0.06 units. You can see this by substituting the two K_{s0}^0 values in (4.19) while holding all others constant.)

The expression for alkalinity in terms of activity coefficients and zero ionic strength constants is obtained analogously. Substitute (4.19) in (4.4), apply (4.11), and rearrange to get

$$A = 2[Ca^{++}] = \left(\frac{2K_{s0}^0 K_{a1}^0 K_H^0}{K_{a2}^0 \gamma_{++} + \gamma_-^2}\,P\right)^{1/3}. \tag{4.21}$$

Here the correction for ionic strength at $P = 1.0$ gives $A = 10^{-1.78}$ (instead of $10^{-1.90}$), and at $P = 10^{-3.5}$ it gives $A = 10^{-3.04}$ (instead of $10^{-3.07}$).   ∎

Of course, the presence of other salts (such as NaCl) can increase the ionic strength, as does the dissolution of $CaCO_3$ in acid (see below). Both γ_{++} and γ_- decrease with increasing I at low ionic strengths (at high ionic strengths, ion pairing becomes important—see Fig. 2.5), and hence you can expect pH and alkalinity to increase with increasing ionic strength in the range $I < 1$.

In the older literature, it is common to see the statement that CO_2 dissolves $CaCO_3$ with the formation of $Ca(HCO_3)_2$. Strictly speaking, there is no such compound as calcium bicarbonate, since it is not possible to crystallize a salt with that composition. However, if only CO_2 and water are present, as you have seen above, the stoichiometry of the solution is given fairly precisely by (4.11) to be $2[Ca^{++}] = [HCO_3^-]$, which is the mass balance you would expect if a hypothetical calcium bicarbonate salt dissolved in water.

The danger in referring to the solution as "calcium bicarbonate" lies primarily in the implication that (4.11) holds when other solutes are present. Addition of acid or base shifts the equilibrium so that this simple stoichiometry no longer holds. A quick look at Fig. 4.1 shows that addition of a small amount of base increases pH and hence makes $[HCO_3^-]$ larger than $2[Ca^{++}]$; addition of a

---

* Recall from (2.36) that $\log \gamma$ is proportional to z^2; hence $\gamma_{++} = \gamma_+^4$ and $\gamma_+ = \gamma_-$ to a first approximation.

small amount of acid makes the pH lower, and hence makes $2[Ca^{++}]$ larger than $[HCO_3^-]$. These effects are considered quantitatively in the next few sections.

DOLOMITE

Dolomite, $CaMg(CO_3)_2$, is another mineral that is almost as important as calcite in limestone rocks. All the above equations remain the same except for the solubility product*

$$K_D = [Ca^{++}][Mg^{++}][CO_3^=]^2 = 10^{-17.0 \pm 0.2} \text{ at } 25°C.$$

Under normal environmental conditions, dissolution of dolomite will yield equal concentrations of Ca and Mg, about $10^{-3.6}$ M (see Problem 27). Another frequent constraint is the simultaneous equilibration of dolomite and calcite with groundwater (see Problem 28), which fixes the ratio $[Ca^{++}]/[Mg^{++}] = K_{s0}^2/K_D$ at approximately 1.0. Note also that at high enough pH, the precipitation of $Mg(OH)_2$ can be important (see Chapter 6).

ADDITION OF STRONG BASE OR STRONG ACID AT CONSTANT P

When a strong base (such as NaOH) or a strong acid (such as HCl) is added to a $CaCO_3$-saturated solution, the charge balance (4.10) is modified by addition of a term in $[Na^+]$ and a term in $[Cl^-]$

$$[Na^+] + 2[Ca^{++}] + [CaOH^+] + [CaHCO_3^+] + [H^+]$$
$$= [HCO_3^-] + 2[CO_3^=] + [OH^-] + [Cl^-]. \tag{4.22}$$

As you can see from Fig. 4.1, $[H^+]$ and the ion pairs are negligible compared to $[Ca^{++}]$ or $[Na^+]$ at pH higher than 8.3, and (unless the partial pressure of CO_2 is very low) $[OH^-]$ is small compared to $[HCO_3^-]$. As NaOH is added, $[HCO_3^-]$ is converted to $[CO_3^=]$, dissolved calcium ion precipitates as $CaCO_3$, and $2[Ca^{++}]$ becomes small compared to $[Na^+]$. In Fig. 4.1, $2[Ca^{++}]$ is less than 10% of $[Na^+]$ (or $[HCO_3^-]$) for pH greater than about 8.8.

    To find the expression for $[Na^+]$ in terms of pH and P, substitute in (4.22) for $[HCO_3^-]$ from (4.4), $[CO_3^=]$ from (4.5), and $[Ca^{++}]$ from (4.6), and neglect the other terms, to get

$$[Na^+] = \frac{K_{a1}K_H P}{[H^+]}\left(1 + \frac{2K_{a2}}{[H^+]}\right) - \frac{2K_{s0}[H^+]^2}{K_{a1}K_{a2}K_H P} + \cdots. \tag{4.23}$$

---

* Holland, H. D. (1978. *The Chemistry of the Atmosphere and Oceans.* New York: Wiley) discusses these problems, and quotes Langmuir, D. (1971. *Geochim. Cosmochim. Acta* 35:1023–1045) for these data. He also gives values for K_D at other temperatures: $pK_D = 16.6$ at 0°C, 16.7 at 10°C, 16.9 at 20°C.

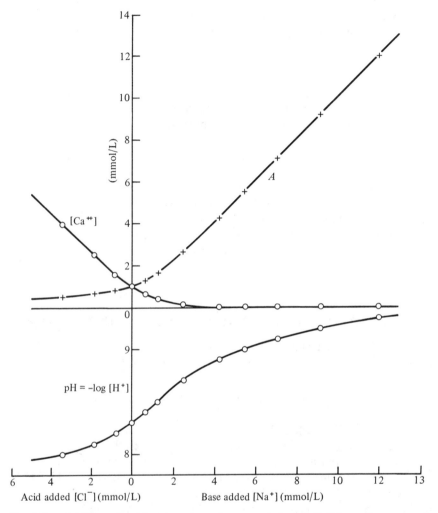

Fig. 4.3. Addition of acid or base to water saturated with $CaCO_3$ under constant CO_2 partial pressure of $10^{-3.5}$ atm. Note the strong buffering of pH, the essentially linear increase in A on the basic branch, and the nearly linear increase in $[Ca^{++}]$ on the acid branch.

If you know the amount of NaOH added, you can calculate $[Na^+]$, and if you also know P, you can solve the polynomial equation in $[H^+]$ to get pH. If you want to make a table or graph of Eq. (4.23), it is computationally simpler to choose a series of values for $[H^+]$ and calculate corresponding values of $[Na^+]$. This procedure is analogous to the alkalinity titration curve calculations for Fig. 3.1. The curve of $[Na^+]$ versus pH, shown in Fig. 4.3, was obtained by substituting in (4.23) some approximate values for the equilibrium constants in dilute solution

$(K_{a1} = 10^{-6.3}, K_{a2} = 10^{-10.3}, K_H = 10^{-1.5}, K_{s0} = 10^{-8.3})$, choosing $P = 10^{-3.5}$, $[Cl^-] = 0$, and assuming pH (that is, $-\log[H^+]$) values from 8.3 to 9.3; $[Ca^{++}]$ can be calculated from (4.6) once $[H^+]$ is known. Note that addition of NaOH in amounts larger than about $3 \cdot 10^{-3}$ M removes more than 90% of $[Ca^{++}]$ (last term of Eq. (4.23)).

At constant CO_2 partial pressure, A is given as a function of $[H^+]$ and P by the substitution of Eqs. (4.4) and (4.5) in (4.16):

$$A = \frac{K_{a1} K_H P}{[H^+]} + \frac{2 K_{a1} K_{a2} K_H P}{[H^+]^2} + \frac{K_w}{[H^+]} - [H^+]. \tag{4.24}$$

The last two terms are normally negligible. Although you could assume a series of pH values and calculate A from Eq. (4.24), it is easier to use the $[Na^+]$ and $[Ca^{++}]$ values calculated above. Combining the definition of carbonate alkalinity (4.16) with the charge balance (4.22), you will find that

$$\begin{aligned} A &= [Na^+] + 2[Ca^{++}] + [CaOH^+] + [CaHCO_3^+] - [Cl^-] \\ &= [Na^+] - [Cl^-] + 2[Ca^{++}] + \cdots. \end{aligned} \tag{4.25}$$

With $[Cl^-] = 0$, this gives the curve of A plotted in Fig. 4.3. It is the same as you would obtain from (4.24). Note that addition of base increases A almost linearly, with only a small curvature near $[Na^+] = 0$ caused by the presence of Ca^{++}.

When a strong acid such as HCl is added to a solution saturated with $CaCO_3$, the carbonate dissolves, increasing $[Ca^{++}]$, as well as the concentration of the total dissolved carbon dioxide. If you have ever dissolved limestone in acid you will remember how vigorously bubbles of CO_2 are evolved when P_{CO_2} exceeds atmospheric pressure.

The left portion of Fig. 4.2 was obtained from a slight generalization of (4.23). Equation (4.22), with the substitution of (4.4), (4.5), and (4.6), or the combination of (4.24) and (4.25), gives:

$$[Cl^-] - [Na^+] = \frac{2 K_{s0} [H^+]^2}{K_{a1} K_{a2} K_H P} - \frac{K_{a1} K_H P}{[H^+]} \left(1 + \frac{2 K_{a2}}{[H^+]} \right) + \cdots. \tag{4.26}$$

(Note $[Na^+] = 0$ for the left side of Fig. 4.3; $[H^+]$, $[OH^-]$, and the ion pairs have been omitted since pH is around 8.) $[Ca^{++}]$ is calculated from (4.6) once $[H^+]$ is known, and A is obtained from (4.25). Since (4.26) is the negative of (4.23), the functions are continuous through the point where $[Na^+] = [Cl^-] = 0$.

LIMITED AMOUNT OF CaCO₃ IN THE PRESENCE OF STRONG ACID

In the previous sections of this chapter, we have assumed an unlimited amount of $CaCO_3$ to be present, so that the solubility product (4.1) will always hold. However, there are situations in lakes, oceans, or industrial processes that involve a limited amount of solid $CaCO_3$. You may be wondering how to proceed with calculations of pH as strong acid is added, when you know that addition of sufficient acid will dissolve all the solid $CaCO_3$.

Heretofore I have not included a mass balance on $[Ca^{++}]$ because of the unspecified amount of Ca^{++} present as solid $CaCO_3$. However, if the number of moles of solid $CaCO_3$ per liter of solution is designated by Q, the mass balance on Ca becomes

$$[Ca^{++}] + Q = C_{Ca}, \tag{4.27}$$

where C_{Ca} is a constant independent of pH or added strong acid or base. So long as $Q \geq 0$, Eqs. (4.1), (4.6), (4.23), (4.24), (4.26), etc., hold. When the last bit of $CaCO_3$ dissolves, however,

$$Q = 0 \quad \text{and} \quad [Ca^{++}] = C_{Ca}. \tag{4.28}$$

The charge balance (4.22) combined with (4.4), (4.5), and (4.28) then gives the titration curve for the unsaturated solution:

$$[Cl^-] - [Na^+] = 2C_{Ca} - \frac{K_{a1}K_H P}{[H^+]}\left(1 + \frac{2K_{a2}}{[H^+]}\right) + [H^+] + \cdots. \tag{4.29}$$

Where the curves (4.26) and (4.29) intersect, the solution is just saturated, but no solid $CaCO_3$ is present. At this point

$$[Ca^{++}] = C_{Ca} = \frac{K_{s0}}{[CO_3^=]} = \frac{K_{s0}[H^+]^2}{K_{a1}K_{a2}K_H P}. \tag{4.6}$$

Given values for C_{Ca}, P, and the equilibrium constants, Eq. (4.6) can be solved for $[H^+]$ at the saturation limit:

$$[H^+] = \left(\frac{C_{Ca}K_{a1}K_{a2}K_H P}{K_{s0}}\right)^{1/2}. \tag{4.30}$$

Then (4.30) can be substituted in (4.29) to eliminate $[H^+]$ and to give explicitly the excess of added strong acid at the saturation limit:

$$[Cl^-] - [Na^+] = 2C_{Ca} - \left(\frac{K_{s0}K_{a1}K_H P}{C_{Ca}K_{a2}}\right)^{1/2}\left[1 + \left(\frac{4K_{s0}K_{a2}}{C_{Ca}K_{a1}K_H P}\right)^{1/2}\right] + \cdots. \tag{4.31}$$

EXAMPLE 3   Take $C_{Ca} = 10^{-2.50}$ and $P = 10^{-3.50}$, with the usual equilibrium constants. (Since the alkalinity with no added acid is 10^{-3} (Fig. 4.3), C_{Ca} must be greater than $0.5 \cdot 10^{-3}$ or the solution will be unsaturated even when there is no added strong acid.) Substituting in (4.30) gives pH = 7.90 for the point where the last bit of $CaCO_3$ dissolves, and (4.29) or (4.31) gives the excess acid as $5.9 \cdot 10^{-3}$ M. On Fig. 4.4, the curve from Fig. 4.3 for saturated $CaCO_3$ solutions has been reproduced, and the point where the last bit of $CaCO_3$ dissolves is marked by a solid circle.   ■

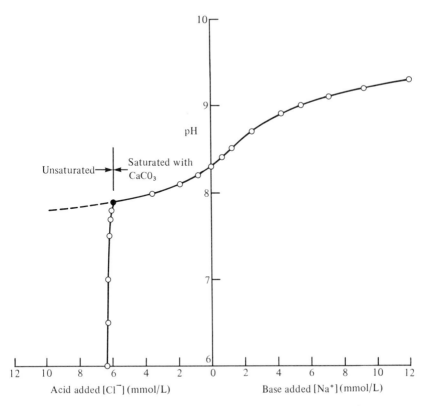

Fig. 4.4.   Addition of acid or base to water saturated with CaCO$_3$ under constant CO$_2$ partial pressure of $10^{-3.5}$ atm. The portion of the pH curve marked "saturated" is the same as in Fig. 4.3. The black dot (intersection with the steeply decreasing curve) corresponds to the point where all the solid CaCO$_3$ present has just dissolved. At pH values below this, the solution has become unsaturated.

How does the pH at saturation depend on C_{Ca} and P? This is given by the solubility product as expressed by Eq. (4.30). In logarithmic form*

$$pH = 4.8 - \frac{1}{2} \log C_{Ca} - \frac{1}{2} \log P - \frac{1}{2} \log \gamma_{++} + \cdots .$$

---

* In the same way that Eqs. (4.19) and (4.21) were obtained, activity coefficients (which are always implicit in the concentration equilibrium constants) can be made explicit in (4.30). The combination of constants for zero ionic strength (from Table 2.2) is:

$$\frac{1}{2} (pK^0_{a1} + pK^0_{a2} + pK^0_H - pK^0_{s0}) = 4.816 \text{ at } 25°C.$$

Most of the activity coefficients cancel, γ_+ is combined with $[H^+]$ to give pH, and all that is left is γ_{++}, which appears in the last term. For the conditions of this example, $I = 10^{-2}$, $\log \gamma_{++} = -0.18$, and the answer changes from pH $= 7.90$ to pH $= 7.91$ (see also Chapter 5, pp. 112–116).

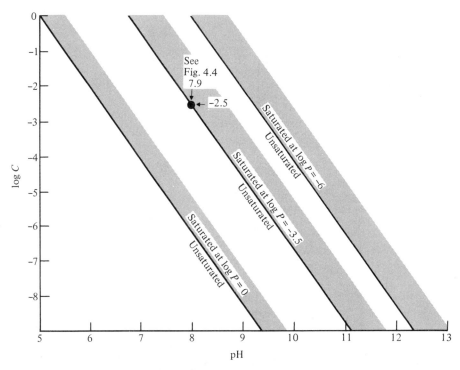

Fig. 4.5.   Lines giving the relationship between calcium ion concentration, pH, and CO_2 partial pressure. For a given value of P, solutions with pH and C_{Ca} to the right of the line are saturated; those to the left are unsaturated. The saturation point of Fig. 4.4 is marked on the line for $P = 10^{-3.5}$.

Figure 4.5 displays a set of curves for C_{Ca} versus pH at different values of P, with the saturation point of Fig. 4.4 marked at $\log P = -3.5$, $\log C_{Ca} = -2.5$, and pH = 7.8. Note that the smaller C_{Ca} and P are, the higher is the pH to which the solution can be raised before $CaCO_3$ precipitates.

When additional acid is added, pH drops steeply, governed by (4.29). Alkalinity is given by (4.25) with $[Ca^{++}]$ replaced by C_{Ca}:

$$A = [Na^+] + 2C_{Ca} - [Cl^-], \tag{4.32}$$

and pH is given indirectly by the usual expression (4.24) for alkalinity. Of course $[H^+]$ will become important in (4.24) or (4.29) when pH reaches 4 or less, but $[OH^-]$ and the ion pairs are normally negligible.

EXAMPLE 4   The unsaturated branch of the curve (Eq. (4.29)) is plotted in Fig. 4.4. Because C_{Ca} is large compared to the terms in $[H^+]$, addition of only $0.40 \cdot 10^{-3}$ mole/L of HCl reduces pH from 7.9 to 6.0.   ■

Of course, all these equations apply equally when NaOH is added to an unsaturated solution containing CO_2 and calcium ion. Until $CaCO_3$ precipitates, (4.29) holds, the saturation point is given by (4.30) and (4.31), and the saturated part of the curve is given by (4.26).

SOLUTION WITHOUT GAS PHASE

If a solution containing only CO_2 and H_2O is in equilibrium with $CaCO_3$, but no gas phase is present, the total carbonate C_T can be varied (by adding or removing CO_2 or carbonate salts from the solution) and this will cause the pH to vary, but not the same as if the partial pressure of CO_2 were changed.

The charge balance is the same as Eq. (4.10) in the discussion at the beginning of this chapter, and as before, a good approximation is usually

$$2[Ca^{++}] = [HCO_3^-] + 2[CO_3^=] + [OH^-] + \cdots, \tag{4.11}$$

with $[Ca^{++}]$ given by (4.1), and the carbonate species given by (2.21) and (2.22).

EXAMPLE 5   Find the pH of a solution with $C_T = 10^{-3}$ in equilibrium with solid $CaCO_3$. Figure 4.6(a) shows the variation of concentrations with pH at constant $C_T = 10^{-3}$. The intersection of the lines representing $2[Ca^{++}]$ and $[HCO_3^-]$ occurs at about pH = 8.3. This is the same value as the constant-pressure example at the beginning of this chapter (Fig. 4.1), because $P = 10^{-3.5}$ gives $[HCO_3^-] = 10^{-3}$ at pH = 8.3. Note that at this pH, both $[CO_3^=]$ and $[OH^-]$ are less than 2% of $[HCO_3^-]$.   ∎

As C_T decreases, $[Ca^{++}]$ increases and pH increases. You can imagine that C_T might get so low that the dominant term on the right-hand side of (4.11) was not $[HCO_3^-]$ but $[CO_3^=]$ or even $[OH^-]$. As long as the solution contains only solid $CaCO_3$, water, and CO_2, however, pH is less than 10.

The highest pH is reached (without addition of strong base) when only $CaCO_3$ and water are present. Then all the carbonate species in solution come from the dissolution of $CaCO_3$, and a mass balance gives

$$[Ca^{++}] = C_T = [CO_3^=] + [HCO_3^-] + [CO_2]. \tag{4.33}$$

(This is the same as $D = 0$, defined on p. 88.) Combination of Eqs. (4.10) (without ion pairs) and (4.33) to eliminate $[Ca^{++}]$ gives a proton condition identical to that for Na_2CO_3 (see Eq. (2.26)):

$$[HCO_3^-] + 2[CO_2] + [H^+] = [OH^-] + \cdots. \tag{4.34}$$

This cannot be solved easily with a diagram like Fig. 4.6, because C_T is not known until the whole problem is solved. An algebraic approach is more direct.

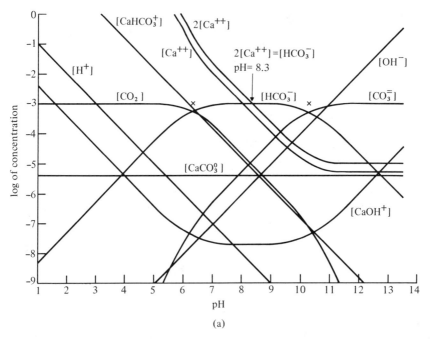

Fig. 4.6.   (a) Logarithmic concentration diagram for $C_T = 10^{-3}$ with lines for $[Ca^{++}]$ and $[CaOH^+]$ corresponding to saturation with $CaCO_3$. Because $CaCO_3$ increases C_T as it dissolves, or decreases C_T as it precipitates, changing pH on this diagram implies concurrent precise removal or addition of CO_2, which is not a realistic constraint. However, the intersection $2[Ca^{++}] = [HCO_3^-]$ for $CaCO_3$ with an equal number of moles of CO_2 gives pH $= 8.3$ (a useful model for saturated river water in limestone regions, as discussed in Chapter 5).

Equations (4.33) and (4.34) can be expressed in terms of $[H^+]$ and $[CO_3^=]$ by using the solubility product (4.1) and the acid–base equilibria (1.1, 2.2, and 2.3), neglecting $[CO_2]$ and $[H^+]$ compared to $[HCO_3^-]$ (note that $[CO_3^=]$ is *not* negligible):

$$K_{s0} = [CO_3^=]^2\left(1 + \frac{[H^+]}{K_{a2}}\right),\tag{4.35}$$

$$[CO_3^=] = \frac{K_w K_{a2}}{[H^+]^2};\tag{4.36}$$

pH can be obtained by substituting (4.36) in (4.35):

$$[H^+]^3 = \frac{K_w^2 K_{a2}}{K_{s0}}\left(1 + \frac{K_{a2}}{[H^+]}\right).\tag{4.37}$$

Equation (4.37) is best solved iteratively. Assuming a starting value of $[H^+] = 10^{-9}$, you can evaluate the term $1 + K_{a2}/[H^+] = 1.05$, and with the

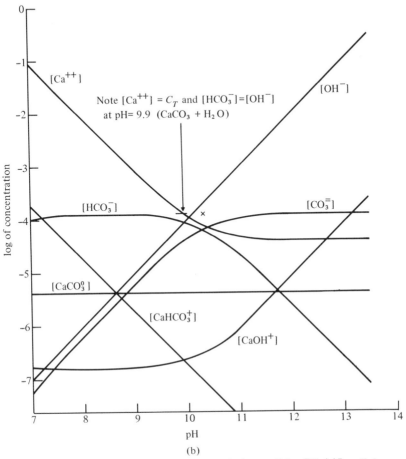

Fig. 4.6.   (b) Portion of the same diagram with the condition $[Ca^{++}] = C_T$ imposed. This corresponds to saturated $CaCO_3$ in H_2O *without* any added CO_2, and gives pH = 9.9.

25°C values of the constants from Table 2.2 you get a first approximation $[H^+] = 10^{-9.9}$. This is used to evaluate the term on the right-hand side, and a second approximation is $10^{-9.89}$. Either (4.35) or (4.36) can then be used to obtain $[CO_3^=] = 10^{-4.55}$. The solubility product $K_{s0} = 10^{-8.52}$ gives $[Ca^{++}] = 10^{-3.97}$, and (4.33) gives $C_T = 10^{-3.98}$. Including activity coefficients and ion pairs gives pH = 9.90, $[CO_3^=] = 10^{-4.50}$ (see problem 21).

Although the logarithmic concentration diagram could not be drawn until all these answers were known, a portion of it is sketched in Fig. 4.6(b). If you had drawn the carbonate lines on one transparent sheet, the calcium lines on another, and moved them up and down until the solubility product (4.1) and the mass balance (4.33) were simultaneously satisfied, then you could have found the intersection of the $[HCO_3^-]$ and $[OH^-]$ lines, which would have satisfied Eq. (4.34). As enthusiastic as I am about logarithmic concentration diagrams, I would not recommend that procedure as a convenient computational method. What Fig. 4.6 *does* show clearly, however, is that the ion pairs $CaOH^+$, $CaHCO_3^+$, and $CaCO_3^0$ were small compared to the principal species in these two examples.

When excess CO_2 is present in the solution, the pH is lower, but C_T (or equivalent information such as A) must be known. Neither (4.33) nor (4.34) holds. If $[CO_3^=]$ and $[OH^-]$ are small compared to $[HCO_3^-]$ in (4.11), substitution of the solubility product gives the simple answer (pH between 7 and 9)

$$[H^+] = \frac{C_T^2 K_{a2}}{2K_{s0}} + \cdots . \tag{4.38}$$

(You will be pleased to note that (4.38) gives pH $= 8.3$ for $C_T = 10^{-3}$, in agreement with Fig. 4.6(a).)

The intermediate region, where neither $[CO_3^=]$ nor $[OH^-]$ is entirely negligible, but $C_T > [Ca^{++}]$, is governed by (4.10) with the ion pairs neglected and the appropriate equilibria substituted:

$$2K_{s0}[H^+]([H^+] + K_{a2})^2$$
$$= C_T K_{a2}[C_T[H^+]([H^+] + 2K_{a2}) + K_w([H^+] + K_{a2})]. \tag{4.39}$$

Knowing C_T and the constants, you can solve this for $[H^+]$ by Newton's method or by successive approximations, using (4.38) as a starting point.

ADDITION OF STRONG ACID OR BASE IN THE PRESENCE OF CaCO₃ WITHOUT GAS PHASE

The familiar diagram of Fig. 4.6 is not as useful in the presence of solid $CaCO_3$ as it might seem. Although the condition $C_T = $ const is easy to maintain in a homogeneous solution containing only CO_2 and strong acid or base (see Chapters 2 and 3), the presence of $CaCO_3$ complicates matters. Strong acid, in addition to shifting the pH to lower values, causes $CaCO_3$ to dissolve and increases $[Ca^{++}]$ and C_T by the amount of $CaCO_3$ dissolved. Similarly, if strong base is added, $CaCO_3$ will precipitate and decrease $[Ca^{++}]$ and C_T by the amount of the precipitate. You might imagine a diagram like Fig. 4.6 with the set of curves for the carbonate species and $[Ca^{++}]$ sliding up and down in synchrony (maintaining $[Ca^{++}][CO_3^=]$ constant and equal to K_{s0}) with the addition of strong acid or base, but it makes more sense to abandon the constraint $C_T = $ const and look for a more appropriate quantity that remains constant.

Since the change in $[Ca^{++}]$ and the change in C_T are constrained to be equal by the stoichiometry of $CaCO_3$, their difference will be independent of whether the solid dissolves or precipitates. Define

$$D = 2(C_T - [Ca^{++}]); \tag{4.40}$$

then, if $\Delta[Ca^{++}] = \Delta C_T$,

$$\Delta D = 2(\Delta C_T - \Delta[Ca^{++}]) = 0.$$

The quantity D is defined with a factor 2 included, to make manipulations with the charge balance simpler:

$$[Na^+] + 2[Ca^{++}] + [CaOH^+] + [CaHCO_3^+] + [H^+]$$
$$= [HCO_3^-] + 2[CO_3^=] + [OH^-] + [Cl^-], \qquad (4.22)$$

$$C_T = [CO_2] + [HCO_3^-] + [CO_3^=]. \qquad \text{(2.18) or (4.41)}$$

Combine (4.22) and (4.41) with (4.40) to obtain:*

$$D_2 = D - [Na^+] + [Cl^-]$$
$$= [HCO_3^-] + 2[CO_2] + [H^+] + [CaOH^+] + [CaHCO_3^+] - [OH^-]. \quad (4.42)$$

Note that the lefthand side of (4.42), like D, is invariant when CaCO₃ is added or removed.

Equations (4.40) and (4.42) yield two expressions for D as a function of C_T and $[H^+]$ by substitution of (4.1), (4.2), (4.9), (2.21), (2.22), and (2.23):

$$D = 2C_T - 2[Ca^{++}] = 2C_T - \frac{2K_{s0}(K_{a1}K_{a2} + K_{a1}[H^+] + [H^+]^2)}{C_T K_{a1} K_{a2}}, \quad (4.43)$$

$$D = [Na^+] - [Cl^-] + C_T \frac{K_{a1}[H^+](1 + K_1^H[Ca^{++}]) + 2[H^+]^2}{K_{a1}K_{a2} + K_{a1}[H^+] + [H^+]^2}$$

$$+ [H^+] + \frac{K_w}{[H^+]}(K_1^{OH}[Ca^{++}] - 1), \qquad (4.44)$$

where $K_1^H = 10^{+1.22}$ is the formation constant for CaHCO₃⁺ (Eq. (4.2)), $K_1^{OH} = 10^{+1.3}$ is the formation constant of CaOH⁺ (Eq. (4.9)), and $[Ca^{++}]$ is given in terms of C_T and $[H^+]$ by the second term of (4.43).

In imagining how you might solve these equations, you should remember that C_T is not constant, but is only a convenient parameter connecting two equations; C_T varies with pH and D. In fact, you could solve (4.44) for C_T and substitute the expression in (4.43) or vice versa to eliminate C_T entirely (each would involve a

---

* D is thus related to the "acidity" defined by the quantity of strong base required to attain a pH equal to that of a Na₂CO₃ solution:

$$D_2 = [Acy] = [HCO_3^-] + 2[CO_2] + [H^+] - [OH^-].$$

As noted in (4.42), this quantity is invariant to addition or removal of CaCO₃ (see Stumm W. and Morgan, J. J. 1981. *Aquatic Chemistry*, 2d ed. New York: Wiley, pp. 186–188). A third related invariant comes from (4.22):

$$[Na^+] - [Cl^-] = A - 2[Ca^{++}] + \cdots,$$

where A is the familiar carbonate alkalinity. This does not reflect the stoichiometry of CaCO₃, however.

quadratic) and obtain one giant equation (see Eq. (4.53), p. 93). The independent variables would then be D and the amount of strong base or acid added ($[Na^+] - [Cl^-]$), and the dependent variable (implicit in the polynomial) would be pH. Other transformations could be made to use A as the intermediate variable instead of C_T.

Before attempting to solve such a cumbersome equation, however, consider only the region near pH = 8.3, where some familiar approximations can be made. Referring to Fig. 4.1 or 4.6, you will see that the ion pairs, $[H^+]$, and $[OH^-]$ are small compared to $[HCO_3^-]$, and you may also recall (p. 23) that when $[HCO_3^-]$ is the dominant carbonate species, $K_{a1}[H^+]$ is much larger than either $K_{a1}K_{a2}$ or $[H^+]^2$. Equations (4.43) and (4.44) then reduce to

$$D = 2C_T - \frac{2K_{s0}[H^+]}{C_T K_{a2}} + \cdots \qquad (\text{pH} \cong 8) \qquad (4.45)$$

and

$$D = [Na^+] - [Cl^-] + C_T + \cdots \qquad (\text{pH} \cong 8). \qquad (4.46)$$

When C_T is eliminated between these two, you have

$$[H^+] = \frac{K_{a2}}{2K_{s0}} (D - [Na^+] + [Cl^-])(D - 2[Na^+] + 2[Cl^-]) + \cdots$$

$$= \frac{K_{a2}}{2K_{s0}} D_2(2D_2 - D) + \cdots \qquad (4.47)$$

Note that if $D = [Na^+] - [Cl^-] = 0$ (that is, pure water and $CaCO_3$ with no added acid, base, or CO_2), Eq. (4.47) gives $[H^+] = 0$, an incorrect answer. (In the last section, p. 87, you saw that the correct answer for this limiting case was approximately pH = 9.9.)

On the other hand, when D is sufficiently large, with $[Na^+] - [Cl^-] = 0$, and pH is between 7 and 9, Eq. (4.46) becomes $D = C_T + \cdots$ and (4.47) becomes

$$[H^+] = \frac{K_{a2}D^2}{2K_{s0}} + \cdots,$$

which is the same as (4.38).

Let $D = 10^{-3}$, and use (4.47) to find how pH varies with addition of base or acid. Choose values for $[Na^+] - [Cl^-]$ in steps of 10^{-4}. The results are plotted as the solid line in Fig. 4.7. (The broken line is obtained by using the full equations derived below.) Two features are worthy of notice. First, pH increases smoothly through the "equivalence point"; there is no inflection at the pH corresponding to a mixture of $CaCO_3$, CO_2, and water, which would compare with the CO_2 or HCO_3^- equivalence points of the alkalinity titration.

Second, the approximations break down when only a little strong base has been added. Equation (4.47) gives $[H^+] = 0$ (infinite pH) when $[Na^+] - [Cl^-] = D/2$, as well as when $[Na^+] - [Cl^-] = D$. Both these results are clearly unrealistic,

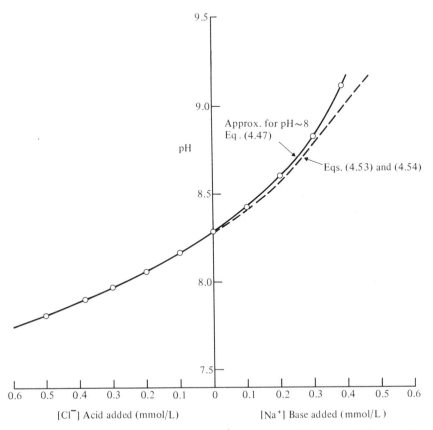

Fig. 4.7. Variation of pH in a solution saturated with $CaCO_3$ ($D = 10^{-3}$) as acid or base is added. The approximation (4.47) and the full equation (4.53) agree for pH $= 8.3$, but diverge as more base is added. The approximation (4.47) gives infinite pH when $[Na^+] - [Cl^-] = D/2 = 5 \cdot 10^{-4}$, which is not realistic, whereas (4.53) gives a smooth increase in pH, leveling out near pH $= 12$ (see Fig. 4.8).

since you ought to be able to add as much strong base as you like and never increase the pH above about 13 or 14. In fact, when the more complete set of equations is solved (see Eq. (4.53) and Fig. 4.8 below), pH does not go up so rapidly as in Fig. 4.7, and never goes to infinity.

EXAMPLE 6   A water with pH $= 7.5$, $[Ca^{++}] = 0.80 \cdot 10^{-3}$, and $C_T = 1.0 \cdot 10^{-3}$ is equilibrated with $CaCO_3$. Find the new pH. Since D is invariant to the addition or removal of $CaCO_3$, it can be calculated from the initial conditions by using (4.40):

$$\frac{1}{2}D = C_T - [Ca^{++}] = 0.20 \cdot 10^{-3}.$$

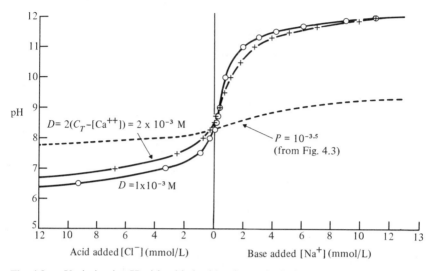

Fig. 4.8. Variation in pH with added acid or base of solutions saturated with $CaCO_3$. Note that increasing D from $1 \cdot 10^{-3}$ to $2 \cdot 10^{-3}$ diminishes the pH range slightly, but changing from constant D to constant P (compare Fig. 4.3) greatly decreases the change in pH because of strong buffering by the gas phase. This same phenomenon is seen in the buffer index diagram of Fig. 4.11.

From (4.42), the other invariant quantity is

$$D_2 = [HCO_3^-] + 2[CO_2] + \cdots = C_T \frac{K_{a1}[H^+] + 2[H^+]^2}{K_{a1}K_{a2} + K_{a1}[H^+] + [H^+]^2} = 1.057 \cdot 10^{-3}.$$

In the final solution, these two equations also hold, but with the final values of C_T, $[Ca^{++}]$, and $[H^+]$. In the final solution, the solubility product also holds:

$$[Ca^{++}] = \frac{K_{s0}}{C_T K_{a1} K_{a2}} (K_{a1}K_{a2} + K_{a1}[H^+] + [H^+]^2).$$

If the final pH is between 7 and 9, $K_{a1}[H^+]$ dominates the polynomial, and simpler equations (analogous to (4.45) and (4.47)) result:

$$C_T = 1.057 \cdot 10^{-3},$$

$$[Ca^{++}] = \frac{K_{s0}[H^+]}{C_T K_{a2}} = C_T - 0.20 \cdot 10^{-3} = 0.857 \cdot 10^{-3},$$

$$[H^+] = (10^{-10.3})(10^{+8.3})(1.057 \cdot 10^{-3})(0.857 \cdot 10^{-3}) = 10^{-8.04},$$

which verifies the assumption. ∎

The general solution of Eqs. (4.43) and (4.44), with fewer approximations so that they will apply over the whole pH range, involves at least a quadratic equation. To simplify manipulations, neglect the ion pairs and let

$$x = D - [Na^+] + [Cl^-] - [H^+] + \frac{K_w}{[H^+]}, \qquad (4.48)$$

$$y = K_{a1}K_{a2} + K_{a1}[H^+] + [H^+]^2, \qquad (4.49)$$

$$z = K_{a1}[H^+] + 2[H^+]^2. \qquad (4.50)$$

Then (4.44) becomes simply

$$C_T = \frac{xy}{z}, \qquad (4.51)$$

and this can be substituted in (4.43) to yield

$$2\frac{x^2 y}{z} - Dx - \frac{2K_{s0}}{K_{a1}K_{a2}}z = 0. \qquad (4.52)$$

The solution of this quadratic in x is

$$x = \frac{z}{4y}\left[D + \left(D^2 + \frac{16K_{s0}}{K_{a1}K_{a2}}y \right)^{1/2} \right]. \qquad (4.53)$$

The easiest procedure for plotting this function is to assume a value for D and a value for pH. This allows you to calculate values for y and z, and to evaluate (4.53) to obtain x. Then x can be used to obtain the amount of excess acid or base by an inversion of (4.48):

$$[Na^+] - [Cl^-] = D - x - [H^+] + \frac{K_w}{[H^+]}. \qquad (4.54)$$

EXAMPLE 7 Substitute $D = 10^{-3}$ and $[H^+] = 10^{-8.3}$ in (4.49), (4.50), and (4.53) to obtain $y = z = 10^{-14.59}$ and $x = 10^{-3.01}$. Putting these values in (4.54), gives $[Na^+] = [Cl^-] = 10^{-8.35}$, or zero within the round-off errors.   ∎

Figure 4.8 displays the effect of added acid and base for two values of D. The curve is qualitatively similar to that obtained with constant P in Fig. 4.3, but the pH change for a given addition of acid or base is much larger. (The curve of Fig. 4.3 is shown as a dotted line on Fig. 4.8 for comparison.) Unlike the approximations used to calculate Fig. 4.7, Eqs. (4.49) to (4.54) yield a curve with an inflection, although this inflection does not occur precisely at zero. Even though there is no reservoir of CO_2 in the gas phase, pH is still limited to the range 6 to 12 when acid or base is added in the amount of as much as 12 times the initial C_T. Of course, this high buffer capacity requires that solid $CaCO_3$ remain.

LIMITED AMOUNT OF $CaCO_3$ AT CONSTANT C_T

Equations (4.51) to (4.53) are derived under the assumption that excess solid $CaCO_3$ is present. Following the same reasoning as for the case when P was held constant, an equation for the pH at which saturation just barely occurs ($Q = 0$) can be obtained from the mass balance on calcium (Eq. (4.28)), from the expression for $[CO_3^=]$ in terms of C_T (2.21), and from the solubility product (4.1). When combined, these three equations give

$$C_{Ca}C_T = \frac{K_{s0}}{K_{a1}K_{a2}}(K_{a1}K_{a2} + K_{a1}[H^+] + [H^+]^2). \tag{4.55}$$

For pH in the range 7 to 9, $K_{a1}[H^+]$ dominates the polynomial, giving the approximate expression

$$\frac{K_{s0}[H^+]}{K_{a2}} = C_{Ca}C_T + \cdots \qquad (pH \cong 8) \tag{4.56}$$

or

$$pH = pK_{a2} - pK_{s0} - \log C_{Ca} - \log C_T + \cdots. \tag{4.57}$$

(Note that (4.56) is equivalent to (4.38) when only $CaCO_3$ and CO_2 are present, that is, $C_{Ca} = C_T/2$.) An equation like (4.57) will appear again in Chapter 6 in the discussion of Langelier's formula (Eq. 6.32).

But C_{Ca} is not a linear function of pH over the whole range. In Fig. 4.9, the full Eq. (4.55) is plotted for several choices of C_T, giving the boundary where saturation with $CaCO_3$ occurs. The linear portion given by (4.56) or (4.57) is marked with a broken line. Comparing the curves in Fig. 4.9 with those in Fig. 4.5, you will see that there is quite a difference between constant partial pressure of CO_2 and constant total carbonate in solution. This is especially noticeable at high pH. With constant P, $CaCO_3$ will precipitate at sufficiently high pH even if C_{Ca} is very low, because there is always enough CO_2 available to increase $[CO_3^=]$ to the point where the solubility product is exceeded.

With constant C_T, on the other hand, increasing pH decreases the solubility limit, but only to the point (pH $\cong$ 11) where all the dissolved carbonate C_T is present as $[CO_3^=]$. Further increase in pH cannot increase C_T, and if C_{Ca} is too low to exceed the solubility product, then $CaCO_3$ does not precipitate, no matter what the pH. For example, if $C_T = 10^{-3}$ and $C_{Ca} = 10^{-6}$, no increase in pH will produce saturation, since the product $[Ca^{++}][CO_3^=]$ never exceeds 10^{-9}, which is less than K_{s0}.

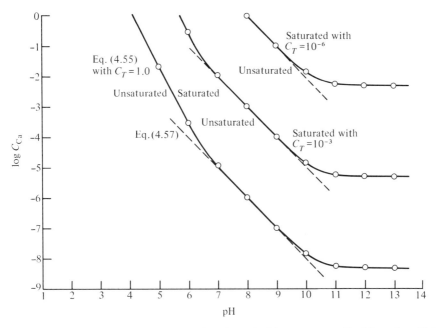

Fig. 4.9.   Solubility of CaCO₃ as a function of pH at constant C_T. Note that, if C_{Ca} is sufficiently low (i.e., $C_{Ca} < 10^{-5.5}$ for $C_T = 10^{-3}$), CaCO₃ will never precipitate no matter how high the pH becomes. This is in contrast to the solubility at constant partial pressure (Fig. 4.5), where sufficiently high pH will always precipitate CaCO₃ even if *P* is very low.

In the unsaturated region, you have already seen that the alkalinity is given by

$$A = [\text{Na}^+] + 2C_{Ca} - [\text{Cl}^-] \tag{4.32}$$

and *A* is related to pH by Eq. (2.28); hence the relationship between added acid or base and pH becomes

$$[\text{Cl}^-] - [\text{Na}^+] = 2C_{Ca} - C_T \frac{K_{a1}[\text{H}^+] + 2K_{a1}K_{a2}}{K_{a1}K_{a2} + K_{a1}[\text{H}^+] + [\text{H}^+]^2} - \frac{K_w}{[\text{H}^+]} + [\text{H}^+]. \tag{4.58}$$

Like Eq. (4.29) (see Fig. 4.4), this gives a curve much steeper in an unsaturated solution than in a saturated solution.

BUFFER INDEX AT CONSTANT *P* WITH CaCO₃ PRESENT

From the previous description of the effects of adding strong acid or base to a solution saturated with calcium carbonate, you should be aware that the presence of solid CaCO₃ can greatly increase the capacity of a carbonate solution to resist changes in pH. Expression of this process in differential form proceeds along the same lines as the derivations of buffer index in Chapter 3.

At constant partial pressure of CO_2, the derivatives are simplest. The buffer index analogous to β_P (Eq. 3.56) is

$$\beta_{P,s} = \left(\frac{\partial[Na^+]}{\partial pH}\right)_{P,s} = -2.303[H^+]\left(\frac{\partial[Na^+]}{\partial[H^+]}\right)_{P,s}, \tag{4.59}$$

where the subscript s indicates the presence of solid $CaCO_3$. The charge balance gives:

$$[Na^+] = [HCO_3^-] + 2[CO_3^=] + [OH^-] - [H^+] - 2[Ca^{++}]$$
$$- [CaHCO_3^+] - [CaOH^+] + [Cl^-]; \tag{4.22}$$

Substitute the equilibrium expressions (4.4) through (4.9), to get $[Na^+]$ as a function of pressure P and $[H^+]$:

$$[Na^+] = \frac{K_{a1}K_H P}{[H^+]} + \frac{2K_{a1}K_{a2}K_H P}{[H^+]^2} + \frac{K_w}{[H^+]} - [H^+]$$
$$- \frac{2K_{so}[H^+]^2}{K_{a1}K_{a2}K_H P} - K_3[H^+] - K_4\frac{[H^+]}{P} - [Cl^-], \tag{4.60}$$

where from (4.7), $K_3 = 10^{+3.03}$ and from (4.9), $K_4 = 10^{-3.07}$. The derivative of (4.60) with respect to $[H^+]$ at constant P (note that $[Cl^-]$ is independent of $[H^+]$ or P, and can also be held constant) gives

$$\left(\frac{\partial[Na^+]}{\partial[H^+]}\right)_{P,s} = -\frac{-K_{a1}K_H P}{[H^+]^2} - \frac{4K_{a1}K_{a2}K_H P}{[H^+]^3} - \frac{K_w}{[H^+]^2}$$
$$- 1 - K_3 - \frac{K_4}{P} - \frac{4K_{so}[H^+]}{K_{a1}K_{a2}K_H P}. \tag{4.61}$$

If this expression is then substituted in (4.59):

$$\beta_{P,s} = 2.303\left(\frac{K_{a1}K_H P}{[H^+]} + \frac{4K_{a1}K_{a2}K_H P}{[H^+]^2} + \frac{K_w}{[H^+]} + [H^+]\right.$$
$$\left. + K_3[H^+] + K_4\frac{[H^+]}{P} + \frac{4K_{so}[H^+]^2}{K_{a1}K_{a2}K_H P}\right) \tag{4.62}$$

and the resulting terms are identified with concentrations of species by means of (4.4) through (4.9), you get

$$\beta_{P,s} = 2.303([HCO_3^-] + 4[CO_3^=] + [OH^-] + [H^+]$$
$$+ [CaHCO_3^+] + [CaOH^+] + 4[Ca^{++}]). \tag{4.63}$$

Compare this with Eq. (3.66)

$$\beta_P = 2.303([HCO_3^-] + 4[CO_3^=] + [OH^-] + [H^+]) \tag{3.66}$$

and you will see that

$$\beta_{P,s} = \beta_P + 2.303(4[\text{Ca}^{++}] + [\text{CaHCO}_3^+] + [\text{CaOH}^+]). \qquad (4.64)$$

As you saw at the beginning of this chapter (Figs. 4.1 and 4.2), the ion pairs are normally small compared to $[\text{Ca}^{++}]$, so that

$$\beta_{P,s} = \beta_P + 9.21[\text{Ca}^{++}] + \cdots. \qquad (4.65)$$

EXAMPLE 8   The graph of Fig. 4.10 compares the two buffer indexes β_P and $\beta_{P,s}$. I have used $P = 10^{-3.5}$ atm and the usual equilibrium constants (compare with Fig. 3.11). Lines giving the concentrations of the major species are also shown. Note that in the alkaline region (pH > 8.3), $[\text{Ca}^{++}]$ falls off rapidly with increasing pH and the buffer index is essentially unaffected by the presence of CaCO₃ ($\beta_{P,s} = \beta_P$). In the acid region (pH < 8.3), however, dissolution of the solid strongly limits the range of pH even when considerable acid is added. This, of course, is no new conclusion; in the form of titration curves, it has been demonstrated in Figs. 4.3 and 4.4.   ∎

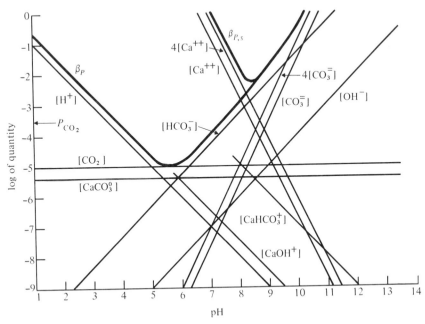

Fig. 4.10.   Buffer index $\beta_{P,s}$ at constant partial pressure of CO_2 ($P = 10^{-3.5}$ atm) with solid CaCO₃ present. For comparison, β_P (Fig. 3.11) has also been plotted. Note the extremely narrow range of pH to which the system is restricted in the presence of the solid.

Note that unless P is as small as 10^{-6}, both $[OH^-]$ and $[H^+]$ are negligible in (4.63). Furthermore, as long as excess solid $CaCO_3$ is present, $\beta_{P,s}$ increases so steeply with decreasing pH that pH is limited to values above 7 no matter how much acid is added. This phenomenon is important for controlling the pH in surface and groundwaters in limestone regions, as well as in other applications discussed in Chapters 5 and 6.

BUFFER INDEX AT CONSTANT C_T WITH $CaCO_3$ PRESENT

Again, by analogy with the derivation of Eq. (3.62) it is straightforward to show (Problem 14) that

$$\beta_{C,s} = \left(\frac{\partial[Na^+]}{\partial pH}\right)_{C_T,s} = 2.303\left[\frac{[HCO_3^-]}{C_T}([CO_3^=] + [CO_2])\right.$$
$$\left. + \frac{2[Ca^{++}]}{C_T}([HCO_3^-] + 2[CO_2]) + [OH^-] + [H^+]\right], \quad \textbf{(4.66)}$$

$$\beta_{C,s} = \beta_C + 4.61[Ca^{++}]\frac{[HCO_3^-] + 2[CO_2]}{C_T}, \quad \textbf{(4.67)}$$

where β_C is given by Eq. (3.62). For simplicity, the ion pairs have been neglected. Although $\beta_{C,s}$ is functionally more complicated than $\beta_{P,s}$, it is essentially the same as β_C in the alkaline region, but increases rapidly with pH below about 8.3 because of the dissolution of $CaCO_3$.

BUFFER INDEX WITH CONSTANT D

As you saw earlier, however, constant C_T is not a very practical constraint when $CaCO_3$ is dissolving or precipitating, but the difference between C_T and $[Ca^{++}]$

$$D = 2(C_T - [Ca^{++}]) \quad \textbf{(4.40)}$$

was invariant under those processes and hence a more useful constraint. Obtaining the appropriate buffer index

$$\beta_{D,s} = \left(\frac{\partial[Na^+]}{\partial pH}\right)_{D,s} \quad \textbf{(4.68)}$$

is, as you might expect, more complicated than the other two buffer index derivations.

The most obvious way to proceed is to differentiate (4.48) with respect to $[H^+]$ holding D and $[Cl^-]$ constant, and use (3.57) to obtain

$$\beta_{D,s} = 2.303\left[[H^+]\left(\frac{\partial x}{\partial[H^+]}\right)_{D,s} + [H^+] + [OH^-]\right]. \quad \textbf{(4.69)}$$

Then obtain the derivative $(\partial x / [H^+])_{D,s}$ from Eqs. (4.49), (4.50), and (4.53) in terms of $[H^+]$ and the equilibrium constants (see Problem 15):

$$[H^+]\left(\frac{\partial x}{\partial [H^+]}\right)_{D,s} = \frac{K_{a1}[H^+](K_{a1}K_{a2} + 4K_{a2}[H^+] + [H^+]^2)}{4y^2}$$

$$\times \left[D + \left(D^2 + \frac{16K_{s0}y}{K_{a1}K_{a2}}\right)^{1/2}\right] - \frac{2K_{s0}}{K_{a1}K_{a2}} \frac{(K_{a1}[H^+] + 2[H^+]^2)^2}{y\left(D^2 + \frac{16K_{s0}y}{K_{a1}K_{a2}}\right)^{1/2}}, \qquad (4.70)$$

where

$$y = K_{a1}K_{a2} + K_{a1}[H^+] + [H^+]^2. \qquad (4.49)$$

EXAMPLE 9   Direct evaluation of $\beta_{D,s}$ from Eqs. (4.69), (4.70), and (4.49), with usual choices of constants and $D = 10^{-3}$ gives the curve shown in Fig. 4.11. In addition, the curves of $[H^+]$, $[OH^-]$, and the two terms of Eq. (4.70) are also plotted. (The sum of these four functions gives $\beta_{D,s}/2.303$.) Note that there is no pH range where $[H^+]$ is a significant part of the sum, and only in the pH range from 9 to 10.5 does the first term of (4.70) contribute significantly. (For higher

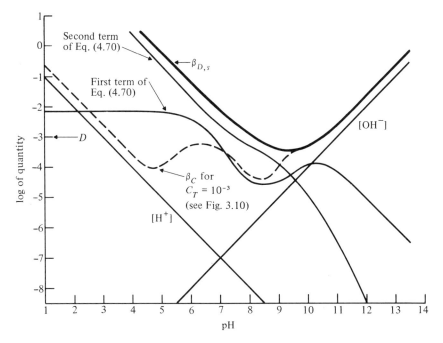

Fig. 4.11.   Buffer index $\beta_{D,s}$ of a closed aqueous system in equilibrium with $CaCO_3$, but without a gas phase. The calculations were made for $D = 10^{-3}$, and the various terms of Eq. (4.70) are also plotted separately. For comparison, the buffer index β_c for a system without $CaCO_3$ has been reproduced from Fig. 3.10 (Eq. (3.62)).

values of D, this first term would become more important, while for lower values of D it would become less important compared to the second term.)

Also plotted is a broken line representing β_C for $C_T = 10^{-3}$, the same as was presented in Fig. 3.10. Only when $[OH^-]$ is dominant, that is, at pH > 10, is β_C the same as $\beta_{D,s}$. As we saw in the case of constant partial pressure, the presence of solid $CaCO_3$ contributes strongly to the buffering action of the carbonate system as pH decreases, although comparison with Fig. 4.10 shows that a solution without gas phase (constant D) is not nearly as strong a buffer as a solution with gas phase CO_2 (constant P).   ∎

By rearranging (4.70), groups of variables can be identified with simple concentrations in solution, but this manipulation is much more complicated than for $\beta_{P,s}$ above (see Problem 16). One form obtainable is

$$\beta_{D,s} = \beta_C + (2.303)(2)[Ca^{++}]\frac{([HCO_3^-] + 2[CO_2])^2}{DC_T}, \qquad (4.71)$$

where

$$\beta_C = 2.303\frac{[HCO_3^-]([CO_3^=] + [CO_2])}{C_T} + [OH^-] + [H^+] \qquad (3.62)$$

is the buffer index at constant total carbonate derived in Chapter 3. Note that when $[Ca^{++}] = 0$, $\beta_{D,s}$ is identical to β_C.

PROBLEMS

1. Find the concentration $[CO_3^=]$ in equilibrium with a solution containing $1.0 \cdot 10^{-3}$ M Ca^{++} and solid calcite.

   Answer: $10^{-5.3}$ M

2. Find the partial pressure of CO_2 in equilibrium with solid calcite if the solution contains $1.0 \cdot 10^{-3}$ M dissolved Ca^{++} at pH $= 8.0$.

   Answer: $10^{-5.1}$ atm

3. Find the total carbonate concentration C_T in equilibrium with solid calcite in a solution containing $1.0 \cdot 10^{-3}$ M dissolved Ca^{++} at pH $= 8.0$. Does the presence of a gas phase affect the answer?

   Answer: $10^{-3.0}$ M; no

4. Rain in equilibrium with atmospheric CO_2 ($10^{-3.5}$ atm) falls on limestone and comes to equilibrium with $CaCO_3$. What is the pH and alkalinity of the runoff water?

   Answer: pH $= 8.3$; $A = 10^{-3.0}$

5. The rainwater in Problem 4, saturated with CO_2 and $CaCO_3$, passes through a limestone-containing soil where microbial activity has boosted the CO_2 partial pressure to 0.2 atm. What is the pH and alkalinity of the water emerging below the soil horizon?

   Answer: pH $= 6.4$; $A = 10^{-1.6}$

6. Suppose the rainwater of Problem 4, saturated with CO_2 and $CaCO_3$, passes through a quartz sand bed that yields no soluble cations, but contains microbially produced CO_2 at a concentration of $[CO_2] = 10^{-2.2}$ M (corresponding to $P = 0.2$ atm). What is the alkalinity and pH of the resulting groundwater?

   Answer: $A = 10^{-1.6}$ M; pH $= 6.9$

7. Acid rain with pH $= 3.0$ falls on a limestone soil. What is the maximum partial pressure of CO_2 in equilibrium with the rain as it first hits the soil? Assume reaction with $CaCO_3$ but not saturation.

   Answer: $P_{max} = 10^{-1.8}$ atm

8. What is the pH and alkalinity of groundwater well below the soil horizon of a limestone soil on which acid rain of pH $= 3$ is falling? Assume saturation with $CaCO_3$.

   Answer: pH $= 8.0$; $A = 10^{-3.0}$.

9. Water saturated with $CaCO_3$ and CO_2 at 10^{-4} atm is treated with NaOH until pH $= 9.0$. What fraction of the alkalinity is due to Ca^{++} at that point?

   Answer: 12%

10. What partial pressure of CO_2 exists in equilibrium with a solution just barely saturated with $CaCO_3$ at pH $= 7.5$ and $C_{Ca} = 10^{-3}$?

    Answer: $10^{-2.2}$ atm

11. If 10^{-3} M HCl is added to the system of Problem 10, at constant P, what pH is reached?

    Answer: pH $= 7.0$

12. If 10^{-3} M HCl is added to the solution of Problem 10, in a closed vessel without a gas phase present, what pH is reached?

    Answer: pH $= 5.96$

13. If the solution in Problem 10 is isolated from the gas phase and excess $CaCO_3$ is added, what pH and alkalinity result?

    Answer: pH $= 7.5$; $A = 2.0 \times 10^{-3}$

14. Derive the expression for $\beta_{C,s}$ given by Eqs. (4.66) and (4.67).

15. Differentiate Eq. (4.53) to obtain $(\partial x/\partial[H^+])_{D,s}$, making use of (4.49) and (4.50). Show that the result is given by (4.70).

16. Combine (4.69) and (4.70) with expressions for $[Ca^{++}]$, $[HCO_3^-]$, $[CO_3^-]$, and C_T to obtain the result given as Eq. (4.71).

17. How are the results of Problem 16 modified by including the ion pairs $CaHCO_3^+$ and $CaOH^+$? Under what conditions would these make a significant contribution to $\beta_{D,s}$?

18. Derive two expressions for D in terms of A and pH, analogous to Eqs. (4.43) and (4.44).

19. The equilibrium constant for the formation of $CaHCO_3^+$ and $CaCO_3^0$ ion pairs at 25°C are given by Eqs. (4.2) and (4.8). The temperature dependence for $CaHCO_3^+$ is given by* $\log K = -2.74 + 0.0133T$ where T is in °K. The constant for $CaCO_3^0$ ion pair formation is given by $\log K = +5.49 - 713/T$. Calculate the alkalinity of water saturated with solid

---

* The 25°C values of Eqs. 4.2. and 4.8 have been combined with temperature coefficients estimated by Jacobson, R. L. and Langmuir, D., 1974. *Geochim. Cosmochim. Acta* 38:301–318.

$CaCO_3$ and gaseous CO_2 (at 1 atm) at $0°C$ and $50°C$. Note: use the appropriate set of K_{s0}^0, K_{a1}^0, and K_{a2}^0 values from Table 2.2.

20. Show that Eq. (4.39) follows from (4.11). Express (4.39) as a series, of which (4.38) is the leading term and those terms following are arranged in order of decreasing magnitude.

21. $CaCO_3$ is shaken with pure CO_2-free water at $25°C$. Find the pH, free carbonate ion concentration, and alkalinity of the equilibrium mixture, including activity coefficients and ion pairs in your calculations. Compare with the results of Eq. (4.37) or (4.39).

    Answer: $pH = 9.90$, $[CO_3^=] = 10^{-4.50}$, $A = 10^{-3.65}$

22. A water containing alkalinity $2.4 \cdot 10^{-3}$ and $[Ca^{++}] = 1.0 \cdot 10^{-3}$, in equilibrium with the atmosphere $(P = 10^{-3.5})$, is isolated from the atmosphere and placed in contact with excess $CaCO_3$. When equilibrium is reached, what are the new values of alkalinity, $[Ca^{++}]$, and pH? Include activity coefficients and ion pairs.

23. The composition of a stream water, a rainwater, and a seawater are given below in millimoles per liter (except for pH):

Component	Stream	Rain	Sea
Alkalinity	+1.11	−0.072	2.38
Ca^{++}	0.50		10.3
Mg^{++}	0.21		53.3
Na^+	0.36	0.009	468.0
K^+	0.03		10.0
NH_4^+		0.016	
Cl^-	0.26	0.012	545.9
$SO_4^=$	0.21	0.028	28.2
$B(OH)_4^- + B(OH)_3$			0.43
SiO_2	0.15		0.01
NO_3^-	0.017	0.026	
pH	7.80	4.14	8.21

Predict the new composition of each if it is allowed to equilibrate with excess $CaCO_3$. See Table 2.1 for equilibrium constants. Assume that rainwater $[CO_2]$ is determined by equilibrium with the atmosphere $(P = 10^{-3.5})$.

24. When samples of water are shipped from distant places, pH may change during shipment, but C_{Ca} and A can be more accurately determined. Some workers estimate the original pH from

$$pH = pK_{a2}^0 - pK_{s0}^0 - \log C_{Ca} - \log A + 2.54 f(I).$$

What limitations does this method have? Distinguish limitations resulting from the basic principle, those resulting from approximations, and those resulting from experimental difficulties.

25. Water is equilibrated with CO_2 at a partial pressure of 10^{-5} atm. The vessel is then sealed and equilibrated with $CaCO_3$. Find the equilibrium pH and alkalinity. Evaluate D.

26. Garrels and Christ (loc. cit., Chapter 3) compare the case where $CaCO_3$ dissolves in pure H_2O with the case where water is equilibrated with atmospheric CO_2, then closed before reaction with $CaCO_3$. This latter case is one simple model for calculating how much carbonate is leached during cave formation. Show that the pH is approximately 9.9 and the dissolved calcium ion concentration $10^{-3.9}$ for both cases.

   Garrels and Christ say: "the role of CO_2 in rainwater probably has been overrated, whereas the effects of hydrolysis and of CO_2 in the soil atmosphere have been underrated." Show that this is so by comparing the case where pure H_2O dissolves $CaCO_3$ with the case where the water is first equilibrated with soil-derived CO_2 at P atm, where $10^{-3.5} < P < 0.2$.

27. The reaction of dolomite with riverwater or groundwater is similar to that of calcite:

$$MgCa(CO_3)_2 + 2CO_2 + 2H_2O \rightleftharpoons Mg^{++} + Ca^{++} + 4HCO_3^-,$$

and its solubility is controlled by a solubility product of the form (see p. 79)

$$K_D^0 = [Ca^{++}][Mg^{++}][CO_3^-]^2 = 10^{-17.0}$$

(at 25°C, zero ionic strength). Show that under the usual conditions,

$$[Ca^{++}] = [Mg^{++}] = \left(\frac{K_{a1}K_H P K_D^{1/2}}{16K_{a2}}\right)^{1/3}.$$

Define the limits of these conditions and evaluate the effect of ionic strength on the calculated Ca^{++} and Mg^{++} concentrations.

28. Water percolating through a rock system containing both calcite and dolomite may come into simultaneous equilibrium with both. Calculate the ratio $[Ca^{++}]/[Mg^{++}]$, given $K_D^0 = 10^{-17.0 \pm 0.2}$ at 25°C, $10^{-16.7}$ at 10°C, and $10^{-16.6}$ at 0°C (Holland, op. cit.), and the solubility product of calcite from Table 2.2. What effect do the partial pressure of CO_2 and the ionic strength have on the ratio?

   Answer: $10^{+0.20 \pm 0.24}$; negligible effect within normal ranges

CHAPTER FIVE

Applications to Geochemistry and Oceanography

Rainwater in uninhabited regions or lakes in high-altitude glaciated areas may be nearly pure water with a little dissolved carbon dioxide; rivers and intermediate lakes contain more solutes derived from the rocks and soil over which the water has passed; saline ponds in desert areas where stream water evaporates may be saturated with minerals containing calcium carbonate, sodium carbonate, calcium sulfate, and sodium chloride. Far exceeding all these varied types of water in total amount, the oceans contain substantial concentrations of solutes but are remarkably uniform in composition throughout the world.

Many geochemical and oceanographic observations can be described, at least partially, by chemical equilibrium models. In this chapter I will focus on those models involving carbon dioxide and its ions. You should bear in mind, however, that there are other equilibrium (as well as nonequilibrium) models necessary to explain many observations, but they are beyond the scope of this book. Therefore, you should be aware of the very extensive geochemical literature.*

In this chapter I will divide natural waters into four groups, based on the principles involved in their equilibrium calculations:

(1) dilute solutions, for which activity coefficients can be accurately predicted using the Davies equation or related theory;

(2) seawater, which is essentially uniform in composition and can be treated as a constant ionic medium;

(3) concentrated solutions, such as are associated with marine evaporites and soda lakes; and

(4) hydrothermal solutions, which may reach temperatures up to 300°C and many thousands of atmospheres of pressure.

---

* Garrels, R. M. and Christ, C. L. 1965. *Solutions, Minerals, and Equilibria.* New York: Harper & Row; Stumm, W. and Morgan, J. J. 1981. *Aquatic Chemistry,* 2d ed. New York: Wiley; Bathurst, R. G. C. 1971. *Carbonate Sediments and Their Diagenesis.* Amsterdam: Elsevier; Riley, J. P. and Skirrow, G. 1975. *Chemical Oceanography.* New York: Academic Press (especially Chapter 9 in volume 2 on CO₂); Holland, H. D. 1978. *The Chemistry of the Atmosphere and Oceans.* New York: Wiley.

RAINWATER

The simplest model of rainwater is pure water saturated with carbon dioxide, and this has been discussed already in Chapter 2 (pp. 18, 19, 24). As you saw in Chapters 2 and 3, introducing activity coefficients makes only minor changes in the calculated pH. The charge balance

$$[H^+] = [HCO_3^-] + \cdots \tag{2.8}$$

remains as before; when combined with the equilibrium constant expressions (2.35) and (2.37), and the assumption that $\gamma_- = \gamma_+$, it yields

$$([H^+]\gamma_+)^2 = K_{a1}^0 K_H^0 P$$

or

$$pH = \frac{1}{2}(pK_{a1}^0 + pK_H^0 - \log P). \tag{5.1}$$

Note that because the activity coefficients of a negative and positive ion are nearly equal in dilute solutions, and the activity coefficient of hydrogen ion is included in the definition of pH, Eq. (5.1) does not contain any explicit activity coefficient values. Ionic strength effects would become apparent only to the extent that γ_+ and γ_- were no longer equal. This effect would begin to be important in the ionic strength range above 0.1 M, but in the absence of other salts, the ionic strength is given by

$$I = \frac{1}{2}([H^+] + [HCO_3^-] + \cdots) = 10^{-pH} + \cdots,$$

and thus $I < 10^{-4}$ for any realistic partial pressure of CO_2. Compare this model with the equivalence point of the alkalinity titration (pp. 49–51), where much higher ionic strength is commonplace.

EXAMPLE 1   Using values for 25°C from Table 2.2, you can obtain pH as a function of P, by means of Eq. (5.1):

$$pH = 3.911 - \frac{1}{2}\log P.$$

Thus, at $P = 10^{-3.50}$ for water in equilibrium with the atmosphere, pH = 5.66. For a soil water with a higher partial pressure of CO_2 ($P = 0.1$ atm), pH = 4.41.

∎

ACID RAIN

Rain in industrialized regions of the world, as well as in areas of volcanic activity, contains acidic solutes other than CO_2. Such precipitation may have pH as low

Table 5.1

**Average composition of rainfall in northeastern
United States (micromoles/liter)***

$[SO_4^=] = 28.1 \pm 3.9$	$[Cl^-] = 12.1 \pm 12.7$
$[NO_3^-] = 25.9 \pm 5.0$	$[Na^+] = 9.3 \pm 12.2$
$[NH_4^+] = 16.0 \pm 4.6$	$[H^+] = 72.3 \pm 12.3$
	$pH = 4.14 \pm 0.07$

* Pack, D. H. 1980. Precipitation chemistry patterns: a
two-network data set. *Science* 208:1143–1145. These
data are from the MAP3S Network.

as 3, and is commonly known as *acid rain*. Eighty-eight monthly average analyses
of rainfall from the northeastern United States during 1978–79 gave the average
composition (and standard deviations) listed in Table 5.1.

The large variability of sodium and chloride ion concentrations can be attributed to the contribution of sea spray. If the coastal stations are eliminated from
the data set, these concentrations become somewhat smaller, but much less
variable:

$$[Cl^-] = 5.6 \pm 1.1 \quad \text{(noncoastal)},$$
$$[Na^+] = 3.2 \pm 0.9 \quad \text{(noncoastal)}.$$

A simple charge balance shows that these six ions probably account for
all of the solutes in the rain. Let $[M]$ represent the sum of all cations and $[X]$
the sum of all anions (each concentration multiplied by its charge, of course)
not accounted for:

$$[Na^+] + [NH_4^+] + [H^+] + [M] = [Cl^-] + [NO_3^-] + 2[SO_4^=] + [X]. \quad (5.2)$$

Using the average values in Table 5.1, you find that $[X] - [M] = 3.4 \ \mu mole/L$,
which is not significantly different from zero.

The ionic strength computed from the average composition is

$$I = \frac{1}{2}([Na^+] + [NH_4^+] + [H^+] + [Cl^-] + [NO_3^-] + 4[SO_4^=]),$$

$$I = 124 \ \mu mole/L = 1.24 \cdot 10^{-4} \ mole/L. \quad (5.3)$$

The Davies equation (2.36) gives $-\log \gamma_+ = 0.006$, or $\gamma_+ = 0.987$.

At pH values near 4, bicarbonate ions form only a very minor part of the
total solutes, and the presence of CO_2 does not affect the mineral acidity when
the partial pressure of CO_2 is typical of the atmosphere. The acidity of the rainwater is given by:

$$A' = [H^+] - [HCO_3^-] + \cdots. \quad (2.17) \text{ or } (5.4)$$

Combining (2.35) and (2.37) gives

$$[HCO_3^-] = K_{a1}^0 K_H^0 P \left(\frac{10^{+pH}}{\gamma_-} \right). \tag{5.5}$$

At higher values of P, $[HCO_3^-]$ may not be entirely negligible compared to $[H^+]$, and substitution of (5.5) in (5.4) yields

$$A' = \frac{10^{-pH}}{\gamma_+} - \frac{(10^{+pH - pK_{a1}^0 - pK_H^0})P}{\gamma_-}. \tag{5.6}$$

EXAMPLE 2 With $P = 10^{-3.5}$ atm, pH = 4.14, $\gamma_- = \gamma_+ = 10^{-0.006}$, and the $25°C$ values of the constants from Table 2.2 ($pK_{a1}^0 = 6.352$ and $pK_H^0 = 1.47$), you get $[HCO_3^-] = 10^{-7.19}$, which is negligible compared to $[H^+]$, so that (5.6) gives $A' = [H^+] = 10^{-4.13}$. ∎

Recall from Chapter 2 that the addition or removal of CO_2 leaves the alkalinity and mineral acidity of a solution unchanged, and you will see that as P increases, pH must decrease so as to leave A' unchanged.

EXAMPLE 3 If P increases to 10^{-1} atm, A' is still $10^{-4.13}$. The other parameters are the same as in the previous example, and Eq. (5.6) yields a quadratic in 10^{+pH} that can be solved with the quadratic formula to obtain pH = 4.05. ∎

It should be clear now that CO_2 alone will decrease, not increase, the pH of rainwater containing mineral acidity. What increases pH? Reaction with bases in the environment. For example, if the acid rainwater were to absorb its high CO_2 content by percolation through soil, it would also tend to dissolve many other components of the soil, including aluminum oxides, iron oxides, calcium carbonate, and salts of organic acids, all of which would tend to raise the pH and buffer it at higher values. These processes lead to river and lake water of pH = 6 to 8.

RIVER WATER

Rainwater (and glacial melt water) reacts with rocks and soil, acquiring solutes as it proceeds on its downhill journey. Some of these reactions are more rapid than others, but clearly the composition of the solutes in river and lake water depend on the history of that water, particularly on the chemical composition of the rocks it has encountered.

Table 5.2
Average composition of river water (10^{-5} mole/L)*

Component	N. America	S. America	Europe	Asia	Africa	Australia	World
HCO_3^-	111	51	156	129	70	52	95.7
$SO_4^=$	21	5.0	25	8.7	14	2.7	11.7
Cl^-	23	14	19	25	34	28	22.0
NO_3^-	1.7	1.2	6.3	1.2	1.4	0.1	1.7
F^-	0.8						0.5
Ca^{++}	53	18	78	46	31	9.8	37.5
Mg^{++}	21	6.2	23	23	16	11	16.9
Na^+	39	17	23	40	48	13	27.4
K^+	3.6	5.1	4.3			3.6	5.9
Fe	0.29	2.5	1.4	0.02	2.3	0.53	1.2
SiO_2	15	20	12	20	39	6.5	21.8
Al	0.9?						<1.5?
Anionic charge	178	76	231	173	133	86	143.3
Cationic charge	191	71	229	178	142	58	142.1
Charge balance	+13	−5	−2	+5	+9	−28	−1.2
Ionic strength	280	102	356	253	199	95	209
$\dfrac{2[Ca^{++}]}{[HCO_3^-]}$	0.95	0.71	1.00	0.71	0.89	0.38	0.78
$-\log P_{CO_2}$ for $CaCO_3$-saturation	3.32	4.41	2.88	3.25	3.93	4.65	3.58

* These data were compiled by Livingstone, D. A. (1963. *Chemical Composition of Rivers and Lakes.* US Geological Survey Professional Paper 440G, Washington DC: US Government Printing Office.). These same data are also quoted by Stumm and Morgan (1970 ed., p. 385) and Holland (op. cit., p. 93) (note errors in the molar value for Mg^{++} in both books). I have converted Livingstone's ppm values to mole/kg (mole/L within roundoff errors) by using atomic weights based on $^{12}C = 12.0000$; Fe and Al are included with silica as uncharged components, because in the normal pH range of 6 to 8 both Fe(III) and Al(III) are probably present principally as colloidal particles. Data for Al and F^- are from Holland (loc. cit.), who also gives a review of other data on trace element concentrations in river water (op. cit., pp. 136–138). Another recent review is by Martin, J. M., and Meybeck, M., 1979. *Mar. Chem.* 7:173–206. (See p. 116 for calculation of last line.)

The composition of river water from six continents, and the world average (weighted by the total river flow in each continent), are given in Table 5.2. Data are available for five anions (of which HCO_3^- always has the highest concentration) four cations, and three common constituents of rocks (Fe, Al, and Si) that are relatively insoluble.

Summing the concentrations of cations (multiplied by the charge on each) and subtracting the sum of anion concentrations (multiplied by charge) gives a charge balance that is essentially zero (-1 out of 140) for the world average river water, and varies from $+13$ for North America to -27 for Australia. Since H^+ is not listed in the table, an extremely acid water might give rise to an apparent negative charge balance (pH $= 4$ would give $10 \cdot 10^{-5}$ mole/L), but it is more likely that the discrepancies reflect uncertainties in the analytical data on which the table is based.

The most pronounced correlation in this set of data is between $[Ca^{++}]$ and $[HCO_3^-]$. As we have already seen, the interaction of solid $CaCO_3$ with dissolved CO_2 produces a ratio $2[Ca^{++}]/[HCO_3^-] \cong 1.0$ (see pp. 75–78). This ratio is calculated for the six continental averages and for the world. It falls between 0.7 and 1.0 (with the exception of Australia, where $[Ca^{++}]$ is much lower than in other parts of the world) and suggests a model for river water that consists of $CaCO_3$, CO_2, and H_2O.

UNSATURATED CaCO₃ AS A MODEL FOR RIVER AND GROUNDWATER

You have already seen a model of rainwater containing only CO_2 and H_2O (Eq. 5.1) as well as a model containing strong acids (Eq. 5.6). When such rain interacts with rocks, it tends to dissolve them, and the simplest model for this process uses a limestone rock consisting of $CaCO_3$. The equations developed in Chapter 4 then apply.

First consider a water in equilibrium with CO_2 at partial pressure P, and let this react with $CaCO_3$, but not to the point where equilibrium is reached. For this case, the solubility product is *not* satisfied, but the charge and mass balances are. Let S be the amount of $CaCO_3$ dissolved per liter of solution; the mass balance on Ca then gives

$$S = [Ca^{++}] + [CaCO_3^0] + [CaHCO_3^+] + [CaOH^+]. \tag{5.7}$$

Let G be the amount of CO_2 dissolved per liter of solution; the mass balance on carbonate then gives

$$G + S = [CO_2] + [HCO_3^-] + [CO_3^=] + [CaCO_3^0] + [CaHCO_3^+]. \tag{5.8}$$

The difference between these two mass balances is independent of S:

$$G + [Ca^{++}] + [CaOH^+] = [CO_2] + [HCO_3^-] + [CO_3^=] = C_T. \tag{5.9}$$

Note that, if you neglect the ion pair $CaOH^+$, $2G$ is equal to D, the invariant defined by Eq. (4.40).

EXAMPLE 4   If the rain contains only CO_2 and H_2O, find G. Here $S = 0$ and the Ca species drop out. Then (5.9), together with (2.35) and (2.37), gives

$$G = C_T^0 = K_H^0 P_0[1 + (K_{a1}^0 10^{+pH_0}/\gamma_-)] + \cdots, \tag{5.10}$$

The initial pH_0 is given by Eq. (5.1) above. At $25°C$ with $P_0 = 10^{-3.50}$, $pH_0 = 5.66$ (see Example 1), and from (5.10), $G = 10^{-4.89}$. This is what you would expect from Fig. 2.1, p. 18.   ■

The other important equation, the familiar charge balance, will also be one of the important equations:

$$[H^+] + 2[Ca^{++}] + [CaOH^+] + [CaHCO_3^+]$$
$$= [HCO_3^-] + 2[CO_3^=] + [X^-] + [OH^-]. \tag{4.10 or 5.11}$$

where $[X^-]$ represents the charge due to the anions of acid rain, such as sulfate, chloride, and nitrate, etc., less the concentration of cations such as Na^+. (See Table 5.1.)

These mass and charge balance equations can be combined with the equilibria of Chapters 2 and 4 to give several variations on the model of how rain becomes unsaturated river water.

First, imagine rain falling on an open limestone surface, maintaining its equilibrium with atmospheric CO_2 as the $CaCO_3$ dissolves. Although G is invariant to $CaCO_3$ dissolution (independent of S), it increases because CO_2 is absorbed from the atmosphere to replace that which reacted with the rock. Whether the rain is acid or not, this will occur; but of course acid rain will dissolve more $CaCO_3$ than a rain consisting only of CO_2 and H_2O. The presence of excess acidity $[X^-]$ affects the charge balance (5.11) but not the mass balance (5.9).

A second variation is to imagine that the rain falls on a limestone soil with a high organic content. This is different only in that the water comes to equilibrium with CO_2 at a higher partial pressure than the atmosphere.

A third model is more complicated. Imagine rain soaking into cracks in limestone where it is able to react with $CaCO_3$ but is no longer in equilibrium with the atmosphere or other source of CO_2. Then G is fixed by the initial conditions (see example above) and remains constant while S increases.

First consider the constant-partial-pressure models. Simplify (5.11) by retaining only the larger terms:

$$[H^+] + 2[Ca^{++}] = [HCO_3^-] + [X^-] + \cdots, \tag{5.11}$$

$$A = 2[Ca^{++}] - [X^-] = \frac{K_H^0 K_{a1}^0 P(10^{+pH})}{\gamma_-} - \frac{10^{-pH}}{\gamma_+}. \tag{5.12}$$

Given P and pH, Eq. (5.12) yields the alkalinity A. Conversely, if you measure alkalinity and pH, you can infer the partial pressure of CO_2 that the water had experienced.

EXAMPLE 5   A stream water has alkalinity 10^{-4} and pH $= 7.0$. What partial pressure of CO_2 would have been in equilibrium with that water? Note from Eq. (4.20) that $I = \frac{3}{2}A + \cdots = 1.5 \cdot 10^{-4}$. The Davies equation (2.36) gives $\gamma_+ = \gamma_- = 10^{-0.006}$. Substitute in (5.12) and solve for P:

$$10^{-4.0} = (10^{-1.47})(10^{-6.352})(P)(10^{+7.00})(10^{+0.006}) - (10^{-7.00})(10^{+0.006}),$$
$$P = 10^{-3.18}.$$

Note that, whereas $[HCO_3^-] = 10^{-4.0}$ and $[Ca^{++}] = 10^{-4.3}$, $[H^+] = 10^{-7.0}$ and could have been neglected. The other terms in (5.11) are also small:

$$[CaOH^+] = 10^{+1.3}[Ca^{++}][OH^-] = 10^{-10.0}, \tag{4.9}*$$
$$[CaHCO_3^+] = 10^{+1.2}[Ca^{++}][HCO_3^-] = 10^{-7.1}, \tag{4.2}$$
$$2[CO_3^=] = 10^{+0.3-10.3}[HCO_3^-]/[H^+] = 10^{-7.0}, \tag{2.3}$$
$$[OH^-] = 10^{-17}/[H^+] = 10^{-7.0}. \quad \blacksquare \tag{1.1}$$

The excess mineral acidity of acid rain makes it more aggressive than CO_2 and H_2O alone. It dissolves more $CaCO_3$, increasing both the alkalinity and (if the water is no longer in equilibrium with the gas phase) the total CO_2 content as well. This effect is controlled by the charge balance:

$$[H^+] + 2[Ca^{++}] = [HCO_3^-] + [X^-] + \cdots. \tag{5.11}$$

If pH and P are kept the same, increasing $[X^-]$ causes $[Ca^{++}]$ to increase (see Problem 11).

Consider now the third model mentioned above, which is also the most complicated one. Rain, "acid" or not, falls into cracks in limestone rock where it is able to dissolve $CaCO_3$ but is no longer able to absorb CO_2 from the atmosphere. As you saw, Eq. (5.9) is independent of S, and will hold regardless of how much $CaCO_3$ dissolves, at least up to the point where the solubility product is satisfied. Similarly, Eq. (5.10) will hold for acid rain as well as pure CO_2-saturated water, except of course that pH_0 will be lower for the acid rain. Thus, you can expect G to be approximately 10^{-5}, since it is determined primarily by $P_0 = 10^{-3.5}$.[†]

---

* Equation (4.9) is written in two lines: $[CaOH^+] = 10^{+1.3}[Ca^{++}][OH^-]$ applies in general; $[CaOH^+] = 10^{-3.07}[H^+]/P$ applies to solutions saturated with $CaCO_3$.

† In a model where rain falls on soil and achieves equilibrium with CO_2 at higher partial pressure than the atmosphere, P_0 is higher ($0.1 - 0.2$ atm) and the water acquires alkalinity from its passage through the soil before entering the cracks where CO_2 is no longer available. In all other respects, the procedure is the same as for this example.

The primary variables now are alkalinity or $[Ca^{++}]$, pH, and C_T. Initial conditions are given by G (and $[X^-]$, if it is significant). Simplifying the equations by retaining only the larger terms results in

$$C_T = [Ca^{++}] + G + \cdots. \tag{5.9}$$

Equation (5.9) can be combined with (5.11) to eliminate $[Ca^{++}]$:

$$[HCO_3^-] + 2[CO_2] + [H^+] + [CaHCO_3^+] = 2G + [X^-] + [OH^-]. \tag{5.13}$$

Retaining only the largest terms in the pH range of 6 to 10 gives

$$[HCO_3^-] + 2[CO_2] = 2G + [X^-] + \cdots. \tag{5.14}$$

$[CO_2]$ and $[HCO_3^-]$ can be expressed in terms of C_T and $[H^+]$ by (2.22) and (2.23). Some simplification results if you note (Fig. 2.3) that the term $K_{a1}K_{a2}$ in the denominator is negligible for pH less than 8 (but be sure to use the full equations for higher pH!). Activity coefficients can be made explicit by using Eq. (2.35) to give C_T as a direct function of pH and the initial conditions:

$$C_T = (2G + [X^-])\frac{(K_{a1}^0\gamma_0/\gamma_-) + 10^{-pH}}{(K_{a1}^0\gamma_0/\gamma_-) + 2(10^{-pH})} + \cdots. \tag{5.15}$$

Equation (5.9) then gives alkalinity as a function of C_T and pH:

$$A = 2[Ca^{++}] - [X^-] = 2(C_T - G) - [X^-] + \cdots.$$

EXAMPLE 6   Rain consisting of water in equilibrium with the atmosphere falls into cracks in limestone. The pH of some of this "crack water" is measured to be 7.0. What are C_T and alkalinity? From Eq. (5.10), $G = 10^{-4.89}$, and, by assumption, $[X^-] = 0$. As above, $\gamma_+ = \gamma_- = 10^{-0.006}$, $\gamma_0 = 1.000$. From (5.15) you can calculate

$$C_T = (10^{+0.301})(10^{-4.89})\frac{10^{-6.352+0.006} + 10^{-7.000}}{10^{-6.346} + (10^{+0.301})(10^{-7.000})},$$

$$C_T = 10^{-4.661}, \qquad A = 2(10^{-4.66} - 10^{-4.89}) = 10^{-4.75}.$$

As in the previous example, you can verify that the terms neglected in (5.9) and (5.11) are all 10^{-7} or smaller and that $[H^+]$ was correctly neglected in (5.14) because of the relatively high pH.   ∎

SATURATED CO_2, $CaCO_3$, AND H_2O AS A MODEL FOR RIVER AND GROUNDWATER

As more $CaCO_3$ dissolves in river or groundwater, equilibrium should eventually be achieved between atmospheric CO_2, other carbonate species, dissolved Ca^{++}, and solid $CaCO_3$. To model this, combine the solubility product ((2.50) or (4.1)), with (5.11) or (5.13), (and (5.9) if appropriate). The simplest approximations of

(5.11) lead to

$$2[Ca^{++}] = [HCO_3^-] + \cdots, \qquad \text{(5.11) or (4.11)}$$

which leads (as in Chapter 4, pp. 75–77) to the standard result for constant P:

$$[H^+]^3 = \frac{K_{a1}^2 K_{a2} K_H^2 P^2}{2K_{s0}}$$

or, with activity coefficients made explicit, to

$$pH = \frac{1}{3}(pK_{a2}^0 - pK_{s0}^0) + \frac{2}{3}(pK_{a1}^0 + pK_H^0 - \log P) + \frac{1}{3}\log 2 + 0.5f(I) + \cdots.$$

$$\text{(4.19) or (5.16)}$$

At ionic strength $1.5 \cdot 10^{-3}$ and $25°C$, this yields* $pH = 8.30$ with $P = 10^{-3.5}$.
    Alkalinity is given approximately by either $[HCO_3^-]$ or $2[Ca^{++}]$:

$$\log A = \frac{1}{3}(pK_{a2}^0 - pK_{a1}^0 - pK_H^0 - pK_{s0}^0 + \log P + \log 2) + f(I). \qquad \text{(4.21) or (5.17)}$$

At ionic strength $1.5 \cdot 10^{-3}$, $25°C$, and $P = 10^{-3.5}$, Eq. (5.17) yields $A = 10^{-3.04}$.
In (5.16) and (5.17), I have used $pK_{s0}^0 = 8.52$. If you use $pK_{s0}^0 = 8.34$, then $\log A$
and pH will both be 0.06 units higher.
    In the case where the rain falls into cracks in the rock and the solution is not
in equilibrium with the atmosphere after the initial conditions, (5.9) applies with
G constant, and $[Ca^{++}]$ is related to pH and C_T through the solubility product

$$[Ca^{++}] = \frac{K_{s0}}{[CO_3^=]}$$

$$= \frac{K_{s0}}{C_T K_{a1} K_{a2}}([H^+]^2 + K_{a1}[H^+] + K_{a1}K_{a2}).$$

$$\text{(4.1) and (2.21)}$$

You saw in Chapter 4 (Eqs. (4.33) to (4.37) and Fig. 4.6(b)) that $CaCO_3$ alone
in water produced a pH of approximately 9.9. Therefore, the first term of the
polynomial can be neglected:

$$[Ca^{++}] = \frac{K_{s0}}{C_T}\left(\frac{[H^+]}{K_{a2}} + 1\right) + \cdots.$$

Substituting in (5.9), we get

$$K_{s0}\left(\frac{[H^+]}{K_{a2}} + 1\right) = C_T(C_T - G) + \cdots. \qquad \text{(5.18)}$$

---

* Equation (4.20) gives $I = \frac{3}{2}A$; A may be also obtained approximately from (4.18) or (5.17) with
$\gamma_{++} = \gamma_- = 1.0$.

In Eq. (5.13), $[CO_2]$ and $[H^+]$ can be neglected to give

$$[HCO_3^-] = 2G + [X^-] + [OH^-] + \cdots. \tag{5.13'}$$

Note that $[OH^-]$ must be retained for the case where G and $[X^-]$ approach zero. As before, $[HCO_3^-]$ can be expressed in terms of C_T and $[H^+]$ (by (2.22)), by neglecting the term $[H^+]^2$ in the polynomial, to obtain*

$$C_T = \left(2G + [X^-] + \frac{K_w}{[H^+]}\right)\left(1 + \frac{K_{a2}}{[H^+]}\right) + \cdots. \tag{5.19}$$

Then C_T as given by (5.19) can be substituted in (5.18) to give a cumbersome polynomial for $[H^+]$:

$$K_{s0}[H^+]^2 - w(w - G)K_{a2}[H^+] - w^2 K_{a2}^2 = 0,$$

and with the activity coefficients made explicit:

$$\frac{K_{s0}^0 10^{-2\text{pH}}}{\gamma_{++}} - w(w - G) K_{a2}^0 \gamma_- 10^{-\text{pH}} - w^2 K_{a2}^{02} \gamma_-^2 = 0, \tag{5.20}$$

where

$$w = 2G + [X^-] + \frac{K_w^0 10^{+\text{pH}}}{\gamma_-}.$$

EXAMPLE 7   What is the highest pH a rainwater with $G = 10^{-4.89}$ and $[X^-] = 0$ can reach on contacting limestone without access to gaseous CO_2? Assume $I = 10^{-4}$ (Examples 5 and 6), so that $\gamma_+ = \gamma_- = 10^{-0.005}$, $\gamma_{++} = \gamma_= = 10^{-0.020}$. Then with $K_{s0}^0 = 10^{-8.52}$, $K_{a2}^0 = 10^{-10.329}$, and $K_w^0 = 10^{-13.999}$, Eq. (5.20) becomes

$$10^{-8.50-2\text{pH}} - w(w - 10^{-4.89})(10^{-10.334-\text{pH}}) - w^2 10^{-20.668} = 0,$$

with

$$w = 10^{-4.589} + 10^{-13.994+\text{pH}}.$$

Eliminating w between the two equations gives a cubic in $10^{-\text{pH}}$ that does not converge easily by successive approximations, so it is best managed by plotting the value of the polynomial as a function of pH and extrapolating to zero, which gives pH $= 9.83$. Equation (5.19) gives $C_T = 10^{-3.91}$ and (5.9) gives $A = 10^{-3.65}$. (Different values for A are obtained from other equations and reflect the approximations used. For example, (5.18) gives $A = 10^{-3.66}$ and (2.28) gives $A = 10^{-3.65}$.)

---

* Note that (5.19) and (5.15) are both approximations to the same equation, in which a term for $[CO_2]$ and the full polynomial of (2.22) appears (compare with Eq. (4.44), p. 89):

$$C_T = \left(2G + [X^-] + \frac{K_w}{[H^+]}\right) \frac{[H^+]^2 + K_{a1}[H^+] + K_{a1}K_{a2}}{2[H^+]^2 + K_{a1}[H^+]}.$$

EXAMPLE 8   Compare the results of the previous example with the pH achieved by a mixture of pure water and $CaCO_3$. This can be calculated by setting $G = 0$ and $[X^-] = 0$ in (5.20) or by making the activity coefficients explicit in (4.37):

$$10^{-3pH} = \frac{K_w^{02} K_{a2}^0}{K_{s0}^0} \frac{\gamma_{++}}{\gamma_-} \left(1 + K_{a2}^0 10^{+pH} \frac{\gamma_-}{\gamma_=}\right). \tag{4.37}$$

In either case, substitution of the above values for constants and activity coefficients (see p. 81 and problem 21 in Chapter 4) gives pH $= 9.90$.*   ∎

Note that the initial equilibration of the water with atmospheric CO_2 produced little effect on the final pH, lowering it by only 0.05 units. If the initial water had been equilibrated with a higher partial pressure of CO_2, or if $[X^-]$ were larger, the resulting pH would be lower. The calculation would have been a little easier, too, if $2G + [X^-]$ were large compared to $[OH^-]$, and $[H^+]$ were large compared to K_{a2}. Then (5.18) and (5.19) could be simplified and combined to give

$$10^{-pH} = \frac{K_{a2}^0}{K_{s0}^0} (\gamma_- \gamma_{++})(2G + [X^-])(G + [X^-]).$$

This equation (compare Eq. (4.47), p. 90) can be used if $2G + [X^-] > 10^{-3.7}$ (which gives pH < 9.3), but it covers only a small fraction of normal environmental conditions.

Higher values of P, such as one might find in water that has passed through soil (where microbial decomposition of organic matter may have produced CO_2 at levels as high as 0.1 or 0.2 atm), will cause increased dissolution of limestone, increasing A and decreasing pH. In the absence of other solutes, however, the ionic strength still remains small. For example, if $P = 10^{-1}$ and $A = 10^{-1.85}$, then pH $= 6.63$ including the activity coefficient corrections. However, these corrections are small ($\gamma_- = 10^{-0.045}$ and $\gamma_{++} = 10^{-0.180}$ at $I = 10^{-2}$) and make a contribution of only 0.05 units to pH and 0.08 units to log A.

The equations of Chapter 4 can be applied equally well to groundwaters (where the condition $P_{CO_2} = $ const is modified to $C_T = $ const; see Problem 12) and cave formation (where the water may again equilibrate with the cave atmosphere; see Problem 15). Such waters may be saturated with calcite or dolomite (see Chapter 4 and Problem 14 of this chapter), or both.

As you have seen, the amount of CO_2 resulting from equilibrium with the atmosphere does not greatly increase the solubility of limestone in water, but the incorporation of higher CO_2 concentrations as the water percolates through soils with higher metabolic activity can substantially increase the aggressiveness of groundwater for dissolving limestone. When such water emerges into a cavern

---

* This calculation agrees with Garrels and Christ (loc. cit.). They also point out that initial equilibration of the water with atmospheric CO_2 (before the calcite is added) makes little difference.

where P_{CO_2} is again close to the atmospheric value ($10^{-2.5}$ to $10^{-3.5}$), release of dissolved CO_2 results in crystallization of $CaCO_3$ in cave formations. In cave atmospheres of less than 100% humidity, evaporation may also assist in precipitation of calcite.

The possibility that two saturated waters may mix and become either supersaturated or unsaturated also has interesting implications for groundwater chemistry and cave geology. This is discussed below (see pp. 152–157).

Using equations similar to the above, you can derive an expression for P_{CO_2} in terms of $[Ca^{++}]$ and $[HCO_3^-]$ and test whether a particular natural water obeys the CO_2–$CaCO_3$ equilibrium model. Combine the solubility product expression (4.1) with the equilibria (2.37) to (2.42) and solve for P:

$$P = \frac{[Ca^{++}] [HCO_3^-]^2 K_{a2}^0 \gamma_-^2 \gamma_{++}}{K_{a1}^0 K_H^0 K_{s0}^0}. \tag{5.18}$$

Using values for 25°C from Table 2.2 ($pK_{s0}^0 = 8.52$) and evaluating the activity coefficients by means of the Davies equation (2.36), you obtain

$$\log P = \log[Ca^{++}] + 2\log[HCO_3^-] + 6.01 - 3.05f(I). \tag{5.19}$$

Results from this calculation are given on p. 108 in the last line of Table 5.2. The world average river data gives $P = 10^{-3.58}$, and the North American data gives $P = 10^{-3.32}$, both close to the average atmospheric partial pressure of CO_2, $P = 10^{-3.5}$. Europe's rivers give a slightly higher $P = 10^{-2.88}$, and Australia and South Amercia give substantially lower values: $P = 10^{-4.65}$ and $10^{-4.41}$, respectively. These differences can most easily be interpreted as expressing the extent to which $[Ca^{++}]$ and $[HCO_3^-]$ are controlled by equilibrium with $CaCO_3$. They confirm that European rivers flow through many limestone-containing areas and receive dissolved carbon dioxide from sources (such as decaying organic matter) in addition to atmospheric CO_2. South America, on the other hand, is dominated by two large rivers (Amazon and Parana–Uruguay) in areas of extremely high rainfall, and although the ratio $2[Ca^{++}]/[HCO_3^-]$ is 0.7, the water is undersaturated with $CaCO_3$ at atmospheric CO_2 pressure.

In spite of the apparently quite different forms of Eqs. (5.12) for unsaturated solutions and (5.16) for saturated solutions, the numerical values of pH obtained are in both cases close to 8.3. This curious coincidence depends, of course, on the partial pressure of CO_2 being near $10^{-3.5}$ atm. The fact that this value *is* the world average atmospheric CO_2 content suggests that the equilibria of calcium carbonate may control (or at least buffer) the atmospheric CO_2 content, a hypothesis that will be considered (pp. 141–149).

OTHER COMPONENTS OF RIVER WATER

River water contains additional solutes besides Ca^{++} and the carbonate species. These include magnesium (from dolomite, magnesium-containing calcites, or magnesium silicate minerals such as talc), sodium, potassium, iron, silica (from

Table 5.3
Derivation of solutes in world average river water* (10^{-5} mole/L)

Solute	Atmosphere	Weathering or solution of minerals						Total
		Silicates	Carbonates	Sulfates	Sulfides	Chlorides	Organic carbon	
HCO_3^-	58		31				7	96
$SO_4^=$	4.5			3.5	3.5			11.5
Cl^-	6					16		22
Ca^{++}	0.5	7	25	3.5		1.5		37.5
Mg^{++}	<0.5	10	6.5			<0.5		16.5–17.5
Na^{++}	5	10				11		26
K^+	<1	5				1		6–7
SiO_2	<1	21						21–22
								−1 to +2
Charge balance	−65	+49	+32	0	−7	0	−7	+2

* After Holland, H. D. (op. cit., p. 140).

feldspars and other igneous rocks), chloride (from halite and other evaporitic salt minerals), sulfate (from gypsum and other sulfate minerals as well as from the oxidation of pyrite and other sulfide minerals), as well as numerous trace elements and organic constituents that are also the results of weathering.

In agricultural areas, fertilizer runoff introduces nitrate, phosphate, potassium, and pesticides; in urban areas, municipal sewage effluents introduce salts, a host of natural and synthetic organic materials, as well as occasional large concentrations of industrially significant elements. It is therefore not possible to make many generalizations about river water, but the average composition of each continent's rivers (Table 5.2) does not vary more than about a factor of three from the world average in most of the major ionic components.*

Table 5.3 summarizes where the major ionic components of river water are derived: from the atmosphere, from weathering any of five classes of minerals, or from organic carbon. As pointed out above, bicarbonate comes primarily from atmospheric CO_2 and its reaction with carbonate minerals such as $CaCO_3$ (also with dolomite and other carbonates), but about 7% comes from the oxidation of organic carbon in sediments.

Carbon dioxide is also an important agent in the weathering of silicate minerals, which produce substantial fractions of the four major cations. A very simple example is albite, a sodium aluminum silicate, which is attacked by CO_2 and water

* Holland (loc. cit.) gives a detailed discussion of the occurrence and sources of the major inorganic components of river water. Table 5.3 summarizes his conclusions.

with the following stoichiometry:

$$NaAlSi_3O_8(s) + CO_2(g) + 5\tfrac{1}{2}H_2O \rightarrow$$

albite

$$Na^+ + HCO_3^- + 2H_4SiO_4 + \tfrac{1}{2}Al_2Si_2O_5(OH)_4(s).$$

kaolinite

This reaction does not necessarily reach equilibrium; but even if it does not, this stoichiometry predicts that the weathering of albite will produce Na^+, HCO_3^-, and dissolved silica in the ratio $1:1:2$.

The equilibrium constant expression for the above reaction[*]

$$\frac{[Na^+][HCO_3^-][H_4SiO_4]^2\gamma_+\gamma_-}{P_{CO_2}} = 10^{-9.7} \tag{5.20}$$

would be satisfied if the water reached equilibrium with both albite and kaolinite as well as atmospheric CO_2. From Table 5.2, you can find the world average river water concentrations for the three dissolved species, and if $P = 10^{-3.5}$, the activity product is a factor of six smaller than the equilibrium constant:

$$\frac{(27.4 \cdot 10^{-5})(95.7 \cdot 10^{-5})(21.8 \cdot 10^{-5})^2(0.951)^2}{(3.16 \cdot 10^{-4})} = 3.55 \cdot 10^{-11} = 10^{-10.45}.$$

Water of this composition, therefore, still can react with albite and atmospheric CO_2 to produce kaolinite and the three dissolved species. You can show that if the CO_2 content of the water were less, the reaction would be closer to equilibrium ($P = 10^{-4.25}$ at equilibrium).

SEAWATER

The solutes in seawater are primarily sodium and chloride ions, with lesser amounts of Mg^{++}, Ca^{++}, K^+, $SO_4^=$, HCO_3^-, and other ions. Table 5.4 compares the chemical composition of "normal" seawater with the average composition of river water. In estuaries and coastal marshland, solutes occur in approximately the same ratios as in seawater, but they are more dilute. Although lakes can vary from nearly pure water to saturated sodium carbonate and calcium sulfate, the ratios of the major ionic components in seawater vary by only a few percent. For example, throughout the Atlantic and Pacific Oceans, the ratio of $[Na^+]$ to $[Mg^{++}]$ varies by only 1.8%, and the ratio of $[Na^+]$ to $[Cl^-]$ varies by only 1.3%.[†]

---

[*] Stumm and Morgan (op. cit. pp. 530–550) give a number of examples of weathering reactions for silicate minerals.

[†] Culkin, F. and Cox, R. A. 1966. *Deep Sea Res.* 13:789–804.

Table 5.4

Composition of normal seawater* and world average river water[†] (mmole/kg)

Component	Seawater	River water
Na^+	468.04	0.274
K^+	10.00	0.059
Mg^{++}	53.27	0.169
Ca^{++}	10.33	0.375
Sr^{++}	0.10	
Cl^-	545.88	0.220
$SO_4^=$	28.20	0.117
Br^-	0.83	
F^-	0.07	0.0053
$HCO_3^-\ (+CO_2 + CO_3^=)$	2.2–2.5[‡]	0.957
$B(OH)_3 + B(OH)_4^-$	0.43	
$Si(OH)_4 + SiO(OH)_3^-$	0.001–0.1[‡]	0.218
$H_2PO_4^- + HPO_4^= + PO_4^\equiv$	0.0001–0.005[‡]	
NO_3^-	0.0001–0.05	0.017
pH	(7.4 to 8.3)	(6.0 to 8.5)
Ionic strength	700	2.09

* After Hansson, I. 1973. *Deep Sea Res.* 20:479–491.

[†] See Table 5.2. The relationship between river water composition and seawater composition is beyond the scope of this book, but is discussed in detail by Holland (op. cit. Chapter 5) in terms of mass balances. This approach has largely supplanted the elegantly simple equilibrium theory of seawater composition put forth by L. G. Sillén (1961 "The Physical Chemistry of Sea Water," in *Oceanography*, M. Sears, ed., Washington D.C.: Amer. Assoc. Advancement Sci. Publ. No. 67. pp. 549–581) and summarized by Stumm and Morgan (op. cit. pp. 572–574). See Problem 53. Another approach to the global mass balance of elements will be found in these papers: Whitfield, M., 1979. *Mar. Chem.* 8:101–123; Whitfield, M., and Turner, D. R., 1979. *Nature* (London) 278:132–137; Turner, D. R., Dickson, A. G., and Whitfield, M., 1980. *Mar. Chem.* 9:211–218.

[‡] Biologically active species vary considerably with time and sampling location.

The total dissolved solids content of seawater is called its salinity, and is usually expressed in parts per thousand by weight (per mille, written ‰). The salinity of normal seawater is approximately 35‰. Although salinity is still almost universally used to describe seawater samples, in these times of automated instrumentation it is more common to measure the electrical conductivity or the chloride ion concentration (by titration with silver ion) and relate these empirically to accurate experiments wherein seawater was evaporated to dryness and weighed. Because the ratio of the various ionic components (except for the biologically active nutrients P, Si, etc., and bicarbonate ion) is nearly constant, the ratio of salinity to chloride ion concentration is also nearly constant. Thus seawater is sometimes described in terms of its chlorinity, which is the weight of chlorine equivalent to the total of chloride, bromide, and iodide in the sample (this is what titration with silver ion measures). The two are related by the empirical expression

$$(\permil \text{ salinity}) = 0.03 + 1.805(\permil \text{ chlorinity}). \tag{5.21}$$

Normal seawater has about 19‰ chlorinity.

In Chapter 2, the ionic strength, which is the principal factor determining the activity coefficients of ionic species, was defined as

$$I = \frac{1}{2}\sum C_i z_i^2, \tag{2.32} \text{ or } \text{(5.22)}$$

where C_i is the concentration of an ion with charge z_i. For the normal seawater of Table 5.4, the species listed* give

$$I = \frac{1}{2}([\text{Na}^+] + [\text{K}^+] + 4[\text{Mg}^{++}] + 4[\text{Ca}^{++}] + 4[\text{Sr}^{++}] + [\text{Cl}^-] + 4[\text{SO}_4^=]$$
$$+ [\text{Br}^-] + [\text{F}^-] + [\text{HCO}_3^-] + 4[\text{CO}_3^=] + [\text{B(OH)}_4^-] + [\text{SiO(OH)}_3^-]$$
$$+ [\text{H}_2\text{PO}_4^-] + 4[\text{HPO}_4^=] + 9[\text{PO}_4^\equiv] + [\text{NO}_3^-] + \cdots). \tag{5.23}$$

You can verify that this leads to $I = 0.700$ mole/kg. The density of seawater at 25°C and 35‰ salinity is about 1.0235 kg/L, and so this corresponds to $I = 0.716$ mole/L. (Since the volume of a seawater sample varies with pressure as well as salinity and temperature, accurate measurements are often reported on a weight basis, i.e., as moles per kilogram of seawater instead of moles per liter. Conversion is straightforward, but requires a table of seawater densities.[†]) Note that K_{a1} and K_{a2} both have units of concentration, and will be 2.35% smaller when expressed in moles per kilogram instead of moles per liter. This corresponds to an increase of pK_a by 0.01 units.

The Davies equation was introduced in Chapter 2 as a simple way to estimate activity coefficients (γ_z) in the absence of more accurate experimental data:

$$-\log \gamma_z = 0.5z^2\left(\frac{I^{1/2}}{1 + I^{1/2}} - 0.2I\right)\left(\frac{298}{t + 273}\right)^{2/3}. \tag{2.36} \text{ or } \text{(5.24)}$$

---

* Note that no ion pairs are listed, although they are known to exist in seawater. The true ionic strength is therefore less than the value given by Eq. (5.23). This is discussed quantitatively later in this chapter (pp. 123–126).
† A recent critical review is by Millero, F. J. Gonzalez, A., and Ward, G. K., 1976. *J. Mar. Res.* 34:61–93. Tables will be found in most treatises on oceanography, such as Riley and Skirrow, op. cit.

With $I = 0.716$ mole/L, at 25°C, this gives $\gamma_1 = 0.69$ for a univalent ion, $\gamma_2 = 0.23$ for a divalent ion. These may be compared with typical activity coefficients derived from experiments*: $\gamma_{Na} = 0.71$, $\gamma_K = 0.63$, $\gamma_{Ca} = 0.26$, $\gamma_{Mg} = 0.28$.

SEAWATER AS A CONSTANT IONIC STRENGTH MEDIUM

These somewhat uncertain activity coefficients make it quite inaccurate to correct equilibrium constants from infinite dilution to the ionic strength of seawater. Furthermore, the numerous weak interactions between ions to form ion pairs (e.g., $NaCO_3^-$, $MgSO_4$) make the full chemical model even more complicated[†]. Fortunately, most of the processes involving acid–base equilibria do not greatly affect the activity of the major ions (Na^+, Mg^{++}, Cl^-, $SO_4^=$, etc.). Since the ionic medium remains essentially constant in composition, the activity coefficients of ions with low concentrations (particularly H^+, Ca^{++}, and $CO_3^=$) also remain constant.

Marine chemists usually use hybrid equilibrium constants with the activity of hydrogen ion but the concentrations of all other ions. For example, the dissociation of CO_2 to H^+ and HCO_3^- is governed by

$$10^{-pH}[HCO_3^-] = K_1'[CO_2]. \qquad \text{(2.39) or (5.25)}$$

Since the seawater is of essentially constant composition, the concentration equilibrium expression would also be constant for variations in pH at constant salinity; the advantage of the hybrid expression (5.25) is that it can use directly those pH values measured by means of a glass electrode calibrated with standard buffers. The advantage of the concentration constants (Hansson scale) is their thermodynamic rigor. They do not depend on any assumptions about single ion activities except that they are constant.

Some hybrid and concentration equilibrium constants for seawater at 25°C and 35‰ salinity are given in Table 5.5.[‡] As you already know from the brief discussion in Chapter 2, they differ substantially from those extrapolated to zero ionic strength and those measured in a noncomplexing medium (e.g., 0.7 M $NaClO_4$). Note especially that the solubility product of calcium carbonate is larger by a factor of 100 in seawater than at zero ionic strength. Variation with salinity is described by equations 5.50 (p. 151), and by Figs. 2.6, 2.7, 2.9, and 5.9.

---

* See, for example, Pytkowicz, R. M. 1975. *Limnol. Oceanogr.* 20:971–975. Note that if you use $I = 0.700$ mole/kg in (5.24), the difference in γ_2 is negligible.

† See Whitfield, M. 1973. *Marine Chem.* 1:251–266; and Table 5.6, p. 124. (Note that the ionic strength calculated from Whitfield's assumed total ionic concentrations (see Table 5.7 and footnote to Table 5.6) is $I_{tot} = 707$ mmole/kg, that is, 1% higher than in Table 5.4.)

‡ Detailed data are given for a wide range of temperatures and salinity by Riley and Skirrow, op. cit., pp. 174–180.

Table 5.5
Hybrid equilibrium constants for sea water at 25°C, 35‰ salinity*

$[H^+][OH^-] = 10^{-13.20}$ $(K_w, \text{mole}^2/L^2)$

$10^{-pH}[OH^-] = 10^{-13.60}$ $(K'_w, \text{mole/L, NBS pH scale})$

$[CO_2] = 10^{-1.536}P_{CO_2}$ $(K_H, \text{mole/L} \cdot \text{atm})$

$[CO_2] = 10^{-1.547}P_{CO_2}$ $(\text{mole/kg} \cdot \text{atm})$

$10^{-pH}[HCO_3^-] = 10^{-6.00}[CO_2]$ $(K'_1, \text{NBS pH scale})$

$[H^+][HCO_3^-] = 10^{-5.857}[CO_2]$ $(K_{a1}, \text{mole/kg, Hansson scale})$

$10^{-pH}[CO_3^=] = 10^{-9.12}[HCO_3^-]$ $(K'_2, \text{NBS pH scale})$

$[H^+][CO_3^=] = 10^{-8.947}[HCO_3^-]$ $(K_{a2}, \text{mole/kg, Hansson scale})$

$10^{-pH}[B(OH)_4^-] = 10^{8.71}[B(OH)_3]$ $(K'_B, \text{NBS pH scale})$

$[H^+][B(OH)_4^-] = 10^{-8.61}[B(OH)_3]$ $(K_B, \text{mole/kg, Hansson scale})$

$10^{-pH}[SiO(OH)_3^-] = 10^{-9.4}[Si(OH)_4]$ (K'_{Si})

$10^{-pH}[H_2PO_4^-] = 10^{-1.61}[H_3PO_4]$ (K_1^P)

$10^{-pH}[HPO_4^=] = 10^{-6.08}[H_2PO_4^-]$ (K_2^P)

$10^{-pH}[PO_4^{\equiv}] = 10^{-8.56}[HPO_4^=]$ (K_3^P)

$10^{-pH}[SO_4^=] = 10^{-1.1}[HSO_4^-]$ (K'_a)

$[Ca^{++}][CO_3^=] = 10^{-6.34}$ $(K_{s0}^{\text{calcite}}, \text{mole}^2/\text{kg}^2)$

$[Ca^{++}][CO_3^=] = 10^{-6.13}$ $(K_{s0}^{\text{aragonite}}, \text{mole}^2/\text{kg}^2)$

* Most data are from Riley and Skirrow (op. cit.): Table 9.16 (after S. E. Ingle et al., 1973. *Mar. Chem.* 1:295–307); Table A9.9 (after C. Mehrbach et al., 1973. *Limnol. Oceanogr.* 18:897–907); Table A9.5 (after J. Lyman, 1956. Ph. D. Thesis, University of California at Los Angeles); Tables A9.6, A9.7, and A9.8 (after I. Hansson, 1973. *Deep-Sea Research* 20:461–491). Silicate constant estimated from 0.5 M NaCl data by J. M. T. M. Gieskes (1974. Chapter 3 in *The Sea*, E. Goldberg, ed. New York: Wiley). Phosphorus data from Kester, D. R., and Pytkowicz, R. M. (1967. *Limnol. Oceanogr.* 12:243–252). Sulfate data from Culberson, C. H., Pytkowicz, R. M., and Hawley, J. E. (1970. *J. Mar. Res.* 28:15–21).

Note that although pH depends on the scale used for calibration, 10^{-pH} does not, strictly speaking, carry any concentration units. On the other hand, since Hansson's "pH" scale is referred to a hydrogen ion concentration in moles/kg of seawater, his constants have been given the units mole/kg. The concentration units of conjugate acid and base (e.g. HCO_3^- and $CO_3^=$) cancel, and so the equilibrium constant has the same value whether these are expressed as moles/L or moles/kg. Where the units make a difference (as in K_H or K_{S0}) conversion can be made using the density of seawater: 1.0235 kg/L.

Note also that ionic concentrations include ion pairs with the major seawater ions of opposite charge, e.g., $[Ca^{++}]$ means $[Ca^{++}] + [CaCl^+] + [CaSO_4] + \cdots$ and $[CO_3^=]$ means $[CO_3^=] + [NaCO_3^-] + [MgCO_3] + \cdots$.

When these constants are used to calculate the equilibrium composition of a typical seawater as pH is varied (without allowing CO_2 to equilibrate with a gas phase, so that C_T is constant except for precipitation of $CaCO_3$), the results can be displayed as a logarithmic concentration diagram such as Fig. 5.1. The complexity arises not only from the additional acid–base systems associated with boron and silicon, but also from the precipitation of $CaCO_3$ and $Mg(OH)_2$ at

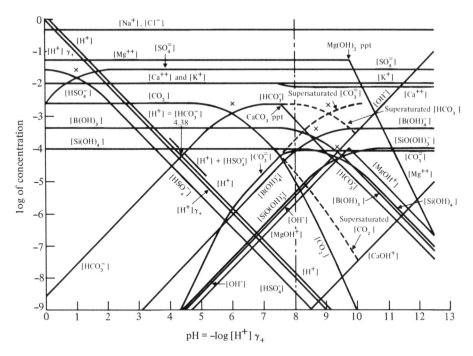

Fig. 5.1. Logarithmic concentration diagram for seawater, illustrating the acid–base behavior of the various components. The vertical line at pH = 8 indicates the composition of open-ocean seawater; the intersection of $[H^+]$ and $[HCO_3^-]$ at pH = 4.38 is the normal end point of the alkalinity titration. The broken lines correspond to solutions supersaturated with calcite; at pH = 9, the difference between saturated and supersaturated solutions is more than a factor of 10 in $[CO_3^-]$ or $[HCO_3^-]$. Most ion pairs (Tables 5.6 and 5.7) are omitted to reduce the complexity of the diagram.

higher pH. Furthermore, the surface waters of the open oceans are normally supersaturated with respect to $CaCO_3$, and to emphasize this the extensions of the curves for constant C_T are shown as dashed lines. Note that complete precipitation (pH > 9) of $CaCO_3$ diminishes $[Ca^{++}]$ only slightly, but reduces C_T from $10^{-2.6}$ to $10^{-4.1}$, that is, by a factor of about 30.

ION PAIRING IN SEAWATER

Ion pairs (except for $CaOH^+$ and $MgOH^+$) have been omitted from Fig. 5.1 but are summarized in Tables 5.6 and 5.7. These data are all results of calculations, and depend on the model used, but the seven models quoted agree qualitatively. For example, about half the sulfate in seawater is present as the free $SO_4^=$ ion, 20%

Table 5.6
Ion-association models for seawater ($298°K$, 1 atm, $35‰$ salinity, pH = 8)*

Anion	Ion pair	\multicolumn{7}{c}{Percentage of anion as ion pair[†]}					
		A	B	C	D	E	F	G
$SO_4^=$	$NaSO_4^-$	18.3	21.9	19.9	37.6	24.7	21.9	37.2
	KSO_4^-	0.6	0.7	0.6	0.4	0.8	0.7	0.4
	$MgSO_4^0$	22.7	23.5	29.0	19.6	24.2	23.5	19.4
	$CaSO_4^0$	3.2	3.7	5.1	2.7	3.9	3.7	4.0
	Free	55.2	50.2	54.5	39.5	46.4	50.2	39.0
CO_3^-	$NaCO_3^-$	17.5	19.1	15.1	17.8	21.4	19.1	17.3
	$MgCO_3^0$	67.0	63.3	68.2	63.0	59.0	63.4	67.3
	$CaCO_3^0$	6.6	7.1	8.4	8.8	7.3	7.1	6.4
	Free	8.9	10.5	8.3	10.4	12.3	10.5	9.0
HCO_3^-	$NaHCO_3^0$	8.6	8.3	8.4	8.0	8.2	8.3	8.6
	$MgHCO_3^+$	17.1	14.5	14.1	9.5	12.4	14.5	17.8
	$CaHCO_3^+$	3.4	3.2	3.1	3.4	2.8	3.2	3.3
	Free	70.9	74.0	74.4	79.1	76.6	74.0	70.3

* From Whitfield, M. 1974. *Limnol. Oceanogr.* 19:235–245. Seawater composition (molalities; Garrels, R. M. and Thompson, M. E. 1962. *Am. J. Sci.* 260:57–66): Na^+ (0.4752), K^+ (0.0100), Mg^{++} (0.0540), Ca^{++} (0.0104), Cl^- (0.5543), $SO_4^=$ (0.0284), $CO_3^=$ (0.000269), HCO_3^- (0.00238).
† Recalculated for the composition shown above with the computer program HALTAFALL (Ingri, N., Kakolowicz, W., Sillén, L. G., and Warnqvist, V. 1967. *Talanta* 14:1261–1286) and the appropriate parameters from the references below: (A) Garrels and Thompson, op. cit.; (B) Berner, R. A. 1971. *Principles of Chemical Sedimentology*. New York: McGraw-Hill; (C) Berner (op. cit.), treating the uncharged ion pairs as dipoles rather than neutral molecules, i.e., with $\gamma_{MgX^0} = 0.8$ (Kester, D. R. 1969. Ph.D. thesis, Oregon State Univ., Corvallis, OR) and $\gamma_{CaX^0} = 0.72$ (Yeatts, L. B. and Marshall, W. L. 1969. *J. Phys. Chem.* 73:81–90); (D) Lafon, G. M. 1969. Ph.D. thesis, Northwestern Univ., Evanston, IL, with the supporting electrolyte convention for defining single-ion activity coefficients; (E) Same as for A, but $\gamma_{\pm(MCl)}$ values used to calculate f_x were calculated from a specific interaction model (Whitfield, M. 1973. *Mar. Chem.* 1:251–256) for the mixture Na^+–K^+–Mg^{++}–Ca^{++}–Cl^-; (F) Calculated from the variable pH model (Whitfield, 1974) at pH 8. The solution composition is as shown in footnote (*), except $C_T = 2.4$ mM and $[H_3BO_3]_T = 0.43$ mM; (G) Kester, D. R. and Pytkowicz, R. M. (1969. *Limnol. Oceanogr.* 14:686–692), with experimental values for sulfate ion-pair formation in a standard seawater solution.

is present as $MgSO_4$ ion pairs, 20% as $NaSO_4^-$ ion pairs, and the remainder as ion pairs with potassium and calcium.

The ionic strength calculated from a typical equilibrium ion-pairing model (Table 5.7) is smaller because part of the charge has been neutralized. The biggest change is the decrease of $SO_4^=$ from 28.4 to 14.2. The ionic strength that takes into account the ion pairs is calculated as follows (note that the uncharged ion pairs

Table 5.7
Composition of typical ion-pair seawater model (mmol/kg)

$[Cl^-] = 554.3$	$[Na^+] = 469.2$	$[K^+] = 9.8$	$[Mg^{++}] = 46.3$	$[Ca^{++}] = 9.2$	$[Cl^-]_T = 554.3$
$[SO_4^=] = 14.2$	$[NaCl] = 0$ (by assumption)	$[KSO_4^-] = 0.2$	$[MgSO_4] = 7.2$	$[CaSO_4] = 1.1$	$[SO_4^=]_T = 28.4$
$[CO_3^=] = 0.03$	$[NaSO_4^-] = 5.7$		$[MgCO_3] = 0.2$	$[CaCO_3] = 0.02$	$[CO_3^=]_T = 0.3$
$[HCO_3^-] = 1.7$	$[NaCO_3^-] = 0.05$		$[MgHCO_3^+] = 0.4$	$[CaHCO_3^+] = 0.08$	$[HCO_3^-]_T = 2.4$
	$[NaHCO_3] = 0.2$	$[K^+]_T = 10.0$	$[Mg^{++}]_T = 54.1$	$[Ca^{++}]_T = 10.4$	
	$[Na^+]_T = 475.2$		$I_{tot} = 707$	$I_{i.p.} = 660$	

This model is consistent with the following formation constants for ion pairs:

$$K = \frac{[MX]}{[M][X]}$$

Species	log K (seawater)	log K^0*	Species	log K (seawater)	log K^0*
$NaSO_4^-$	−0.1	+0.72	$MgHCO_3^+$	+0.7	+1.16
$NaCO_3^-$	+0.6	+1.27	$MgOH^+$	+1.9	+2.58
$NaHCO_3$	−0.6	−0.25	$CaSO_4$	+0.9	+2.31
KSO_4^-	+0.2	+0.96	$CaCO_3$	+1.9	+3.2
$MgSO_4$	+1.0	+2.36	$CaHCO_3^+$	+0.7	+1.26
$MgCO_3$	+2.1	+3.4	$CaOH^+$	+0.6	+1.30

* From Garrels and Christ (op. cit., p. 96). See also problem 56, p. 182, for a more recent set of constants.

such as $MgCO_3$ do not appear in the equation):

$$I_{i.p.} = \frac{1}{2}([Na^+] + [K^+] + 4[Mg^{++}] + 4[Ca^{++}] + [Cl^-]$$
$$+ 4[SO_4^=] + [NaSO_4^-] + [KSO_4^-] + 4[CO_3^=]$$
$$+ [NaCO_3^-] + [HCO_3^-] + [MgHCO_3^+] + [CaHCO_3^+])$$
$$= 600 \text{ mmole/kg}. \tag{5.26}$$

Is this 7% decrease in ionic strength of great significance? No. If you refer to Figs. 2.6 and 2.7 (pp. 35–37), which show the ionic strength dependence of K_{a1} and K_{a2}, you will see that the minimum in the curve of pK versus ionic strength comes at approximately $I = 0.7$. The difference in the pK values at $I = 0.66$ and $I = 0.70$ is less than 0.01 logarithmic unit.

The major effect of ion pairing occurs, therefore, not through the small decrease it produces in ionic strength, but in its removal of important ions such as $CO_3^=$ and HCO_3^- from their acid–base equilibria. I have already alluded to this in Chapter 2, when I pointed out that concentration equilibrium constants measured in seawater (Hansson scale) were significantly different from those measured in noncomplexing media of the same ionic strength. The ion-pairing model whose results are displayed in Table 5.7 can be used to show that the difference between seawater and noncomplexing media is quantitatively explainable by ion-pairing effects.

Consider the first acidity constants:

$$K_{a1} = \frac{[H^+][HCO_3^-]_f}{[CO_2]}. \tag{5.27}$$

In a noncomplexing medium such as $NaClO_4$, where the interactions with the medium ions are of approximately the same strength as the interactions with solvent molecules, the three concentrations in Eq. (5.27) are considered to be "free" concentrations (subscript "f"). In seawater, on the other hand, free HCO_3^- accounts for only $1.7 \cdot 10^{-3}$ out of $2.4 \cdot 10^{-3}$ mole/kg; the rest is present as ion pairs with Na^+, Mg^{++}, and Ca^{++}. This can be expressed as

$$K_{a1}^T = \frac{[H^+][HCO_3^-]_T}{[CO_2]}, \tag{5.28}$$

where

$$[HCO_3^-]_T = [HCO_3^-]_f + [NaHCO_3] + [MgHCO_3^+] + [CaHCO_3^+].$$

According to the model of Table 5.7, neither H^+ nor CO_2 forms any ion pairs, so these are the same in (5.27) and (5.28). If you substitute $[HCO_3^-]_f = 1.7 \cdot 10^{-3}$ and $[HCO_3^-]_T = 2.4 \cdot 10^{-3}$, you can calculate

$$pK_{a1} - pK_{a1}^T = \log \frac{2.4}{1.7} = +0.15.$$

Now refer to Fig. 2.6 (p. 35). At $I = 0.66$, the Davies equation gives $pK_{a1} = 5.97$ to 6.01, depending on the assumption made for γ_0. This is in agreement with the experimental values obtained in noncomplexing KNO_3 media. Converted to moles per kilogram, these values are 5.98 to 6.02. The value for seawater from Table 5.5 (Hansson scale) is $pK_{a1}^T = 5.857$. Thus the experimental difference $pK_{a1} - pK_{a1}^T$ is 0.12 to 0.16, in good agreement with the calculations above based on the model of Table 5.7.

A similar argument can be made for the second acidity constant, where the effect is larger. By analogy with the above:

$$K_{a2} = \frac{[H^+][CO_3^=]_f}{[HCO_3^-]_f} = \frac{[H^+](0.03)}{(1.7)}, \tag{5.29}$$

$$K_{a2}^T = \frac{[H^+][CO_3^=]_T}{[HCO_3^-]_T} = \frac{[H^+](0.30)}{(2.4)}. \tag{5.30}$$

The numerical values (in mmoles per kilogram) were taken from Table 5.7. Thus the ion-pairing model predicts a difference in pK values of

$$pK_{a2} - pK_{a2}^T = \log \frac{(1.7)(0.3)}{(0.03)(2.4)} = +0.85.$$

The experimental data or the Davies equation curve in Fig. 2.7 (p. 37) gives (in mmoles per kilogram) $pK_{a2} = 9.71$, whereas the Hansson scale value for seawater (Table 5.5) is $pK_a^T = 8.95$. The experimental difference is thus $+0.76$, in agreement with the model.

As you already know, the biggest effect of ion pairing is on the solubility of $CaCO_3$. Note that

$$[Ca^{++}][CO_3^=] = (9.2 \cdot 10^{-3})(0.03 \cdot 10^{-3}) = 10^{-6.56}, \tag{5.31}$$

$$[Ca^{++}]_T[CO_3^=]_T = (10.4 \cdot 10^{-3})(0.3 \cdot 10^{-3}) = 10^{-5.51}. \tag{5.32}$$

The numerical values were taken from Table 5.7. Comparing these values with K_{s0} from Fig. 2.9 (p. 40), you will find the model seawater supersaturated with $CaCO_3$. From the upper Davies equation curve in Fig. 2.9 at $I = 0.66$, you find $pK_{s0} = 7.25$ (mole/L)^2 or 7.27 (mole/kg)^2, not 6.56; and from the seawater data (lower curve) you would have expected $pK_{s0}^T = 6.12$ to 6.34, not 5.51.

However, the differences, which reflect the extent to which the ion-pairing model explains the empirically observed data, are in agreement. The Table 5.7 model, expressed in Eqs. (5.31) and (5.32), gives

$$-\log[Ca^{++}][CO_3^=] + \log[Ca^{++}]_T[CO_3^=]_T = 1.05.$$

From Fig. 2.9 data $pK_{s0} - pK_{s0}^T = 0.93$ to 1.15.

Does ion pairing have any effect on the alkalinity titration? Very little. As HCl is added to seawater, it reacts not only with HCO_3^- and $CO_3^=$ but also with the ion pairs containing the carbonate species. Thus the equivalence points remain the same, although the precise shape of the titration curve need not. The effect on ionic strength is extremely small. For every HCO_3^- (or ion pair containing HCO_3^-) reacted to form CO_2, one Cl^- is added. Reaction of $CO_3^=$ (or ion pair containing $CO_3^=$) requires 2 moles of HCl per mole of $CO_3^=$, and introduces two negatively charged ions of Cl^- per $CO_3^=$ reacted. The overall effect is to increase the ionic strength of the model in Table 5.7 from 660 to 662 mmole/kg. Furthermore, the quantity of cations released from ion pairs (as HCO_3^- and $CO_3^=$ react) which could affect the acidity constants (K_{a1}^T and K_{a2}^T), and hence the shape of the titration curve, is also small compared to their total concentration in seawater (see Problem 30). From data in Table 5.7, you can show that total reaction of the carbonate species increases $[Na^+]$ by 0.04%, leaves $[K^+]$ unchanged, increases $[Mg^{++}]$ by 1.3%, and increases $[Ca^{++}]$ by about 10%. Only this latter effect could be significant.

TOTAL ALKALINITY AND CARBONATE ALKALINITY IN SEAWATER

When seawater is titrated with standard HCl to the CO_2 end point (see Figs. 3.6 and 3.7), the total alkalinity or *titration alkalinity* A_T is obtained. This is usually converted to carbonate alkalinity by calculating the concentrations of relevant weak acids and bases. Recall (pp. 51–52) that

$$A_T = A_C + \sum_i [HA]_i' + \sum [B]_i^0 + \cdots, \qquad \text{(3.35) or (5.33)}$$

where

$$A_C = [HCO_3^-]_T + 2[CO_3^=]_T + [OH^-] - [H^+]. \qquad (3.1)$$

Here $[HA]_i'$ are various weak acids (e.g., HSO_4^-) present at the equivalence point and $[B]_i^0$ are various weak bases (e.g., $B(OH)_4^-$) present at the start of the titration; for example (recall Eqs. (3.51) to (3.53), pp. 63–64.)

$$A_T = A_C + \frac{K_B' C_B}{K_B' + 10^{-pH}} + \cdots \qquad (5.34)$$

where C_B is the total boron content of seawater, and K_B' is the acidity constant for boric acid ($10^{-8.71}$ at 25°C and 35‰ salinity; see Table 5.5, also Eq. 5.50 (pp. 150–151). Note that in the definition of A_C, $[HCO_3^-]_T$ and $[CO_3^=]_T$ include the ion pairs with Na^+, Ca^{++}, Mg^{++}, etc.

In a typical seawater, 99.2% of A_T can be accounted for by carbonate and borate. At pH = 8, these are 89.6% $[HCO_3^-]_T$, 6.7% $[CO_3^=]_T$, and 2.9% $[B(OH)_4^-]$, respectively.* Silicate $[SiO(OH)_3^-]$ accounts for about 0.2% of A_T, $[HPO_4^=]$

---

* Edmond, J. M. 1970. *Deep Sea Res.* 17:737–750.

for 0.1%, and $[PO_4^=]$ for 0.05%, depending on the total concentrations of these nutrients, which are highly variable (see pp. 63–64; also Problems 36 and 37).

The magnesium hydroxide cation $[MgOH^+]$ accounts for about the same amount of alkalinity (0.1% at pH = 8) as does $[OH^-]$ (see Problem 39). The only weak acid of importance affecting the equivalence point is $[HSO_4^-]$. Since $pK_a' = 1.1$ in seawater, and $C_S = 28.2 \cdot 10^{-3}$, then at pH = 4.5

$$[HSO_4^-]' = \frac{10^{-pH}C_S}{10^{-1.1} + 10^{-pH}} = 1.12 \cdot 10^{-5} \text{ mole/kg}, \tag{1.15}$$

or 0.4% of the total alkalinity ($2.67 \cdot 10^{-3}$). Since this contribution of $[HSO_4^-]$ is almost exactly the same for all seawaters, it has been the convention for oceanographers to ignore this effect and include it in A_c. Such an attitude is certainly not consistent with the frequent publication of equations for total alkalinity that include phosphate species, NH_4^+, $MgOH^+$, and F^-, as well as borate and silicate.

With a total fluoride concentration of $5.2 \cdot 10^{-5}$, and $pK_a' = 2.5$, you can calculate (as above) that $[HF]' = 5.2 \cdot 10^{-7}$ and that this contributes about 0.02% to A_T. Similarly, $[H_3PO_4]'$ contributes less than 0.001% to A_T and is justifiably ignored.

In anoxic waters, where substantial concentrations of sulfide and ammonia can be present, it is essential to know these concentrations and to include them in the total alkalinity:

$$A_T = A_C + [HS^-]^0 + 2[S^=]^0 + [NH_3]^0, \tag{5.35}$$

where

$$[HS^-]^0 = \frac{10^{-pH^0}C_{H_2S}}{10^{-pH^0} + 10^{-6.7}} + \cdots, \qquad [NH_3]^0 = \frac{10^{-9.2}C_{NH_3}}{10^{-pH^0} + 10^{-9.2}},$$

and pH^0 is the value at the start of the titration. Since pK_{a2} for H_2S is of the order of 12 or 13, $[S^=]^0$ will only be important in highly alkaline waters.

FIELD OBSERVATIONS OF SEAWATER

The usual carbonate-related quantities measured on samples of water collected at sea include pH, total CO_2 in solution, total alkalinity, and partial pressure of CO_2 in equilibrium with the solution. Carbonate alkalinity (A_c), defined by Eq. (5.34), is normally calculated from total alkalinity* (A_T) by Eq. (5.33). As discussed above, the principal correction required is due to borate. If the equilibrium constants are known, any two of the variables A_c, C_T, pH, P_{CO_2} will suffice to determine the others; conversely, a measurement of the three independent

---

* In the oceanographic literature, total alkalinity is sometimes represented by TA or simply A; carbonate alkalinity by CA; and total carbonate by TCO_2 or $\sum CO_2$.

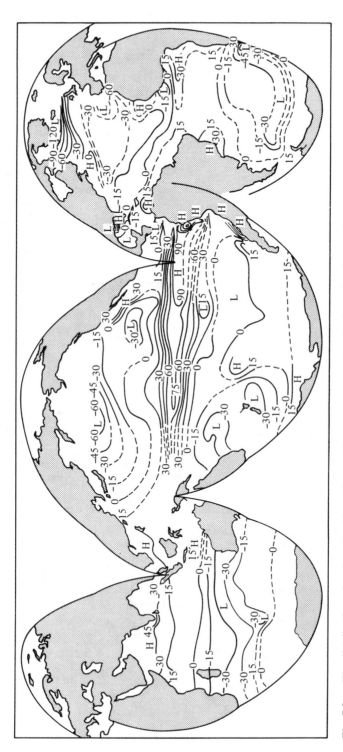

Fig. 5.2.   The distribution of carbon dioxide in the world oceans expressed as the departure in ppm from equilibrium with atmospheric CO_2. Here H indicates high values and L—low values. (From Keeling, C. D. 1968. *J. Geophys. Res. 73*:4547.)

quantities pH, C_T, and A_c will suffice to determine K'_1 for each water sample. Measurement of P_{CO_2} in addition permits an independent estimate of the Henry's Law constant for each sample.

Surface ocean waters range in pH from about 7.8 to 8.4, in total carbonate (C_T) from 2.0 to $2.2 \cdot 10^{-3}$ mole/kg, in equilibrium partial pressure of CO_2 (P_{CO_2}) from less than 0.25 to more than $0.45 \cdot 10^{-3}$ atm. Alkalinity varies more or less directly with salinity, and is sometimes expressed as *specific alkalinity* or total alkalinity per unit of chlorinity. This quantity ranges from 0.12 to $0.13 \cdot 10^{-3}$ equivalent/liter (Cl‰), which corresponds to $A_T = 2.3$ to $2.6 \cdot 10^{-3}$ mole/kg for normal seawater.

Photosynthesis by planktonic algae consumes CO_2 in the surface waters:

$$CO_2 + H_2O \rightarrow \text{``CH}_2\text{O''} + O_2,$$

and this primary production causes large areas of the open ocean to be under-saturated with CO_2 (see Fig. 5.2), but upwelling of deep waters in the equatorial region and along the west coast of the American continent brings water super-saturated with CO_2 to the surface. This CO_2 is formed when organic matter decomposes in the deeper waters of the ocean.

One of the more unusual field studies of carbonate system parameters was the profile taken by a team of eight investigators[*] in 1969 at a station in the Eastern Pacific ($28°20'$N, $121°41'$W) as part of the National Science Foundation GEOSECS (Geochemical Ocean Section) program. All four parameters (pH, P_{CO_2}, A, C_T), as well as temperature and salinity, were measured to a depth of over 4000 m.

The alkalinity values measured by Edmond by using Gran's potentiometric titration method (as described on p. 61), agreed within 1% with values measured by Culberson by a pH method (Fig. 5.3; see also Problems 28 and 29).

Edmond measured total carbonate by using the difference between the first and second end points of the Gran titration. His values were less than 1% lower than those calculated from Culberson's alkalinity and pH data (Fig. 5.4), but about 2% higher than the measurements made by Weiss by using a chromato-graphic technique to separate the gases stripped from the acidified sample by a hydrogen or helium carrier gas. After the cruise was completed, a number of samples were analyzed at land-based laboratories with an infrared technique. These latter results were 3 to 5% higher than those Weiss obtained on shipboard, and the discrepancy was determined to be the result of microbial decomposition of organic matter in the samples during the three-month storage period.

The most curious inconsistency in this careful study was that, when two of the four measured parameters were used to calculate the remaining two, calculated values often differed systematically from the measured values.[†] This discrepancy could be eliminated if the second dissociation constant K'_2 were increased to

---

[*] Takahashi, T., Weiss, R. F., Culberson, C. H., Edmond, J. M., Hammond, D. E., Wong, C. S., Li, Y.-H., and Bainbridge, A. E. 1970. *J. Geophys. Res.* 75:7648–7666.

[†] See Skirrow, G. 1975. *Chemical Oceanography*, vol. 2. New York: Academic Press, Chapter 9, pp. 160–169, for full discussion.

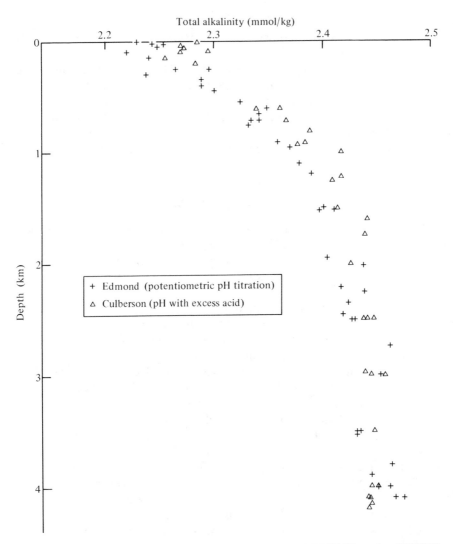

Fig. 5.3.   Total alkalinity as a function of depth at the GEOSECS station (28°20′N, 121°41′W) determined by two methods. Edmond used the Gran potentiometric pH titration; Culberson added excess acid, purged carbon dioxide from the solution, and measured the final pH.

$1.06 \cdot 10^{-9}$; this is 30% more than the then accepted values obtained by Lyman (1956). But this suggested change in equilibrium constant was *not* supported by later work. For example, Lyman gave $K_2' = 8.13 \cdot 10^{-10}$ at 25°C and 35‰ salinity; Mehrbach used a different pH scale, but when his value was corrected to Lyman's pH scale, he got $8.12 \cdot 10^{-10}$, agreeing within 0.12%. Hansson's results, again

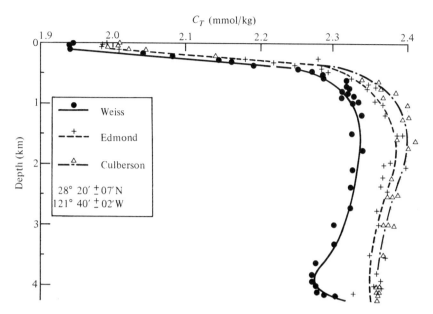

Fig. 5.4. Total dissolved carbonate at the GEOSECS station. Weiss used a chromatographic technique; Edmond and Culberson calculated C_T from alkalinity (see Fig. 5.3) and pH. (From Skirrow, G. 1975. In *Chemical Oceanography*, J. P. Riley and G. Skirrow, eds., vol. 2. New York: Academic Press, Chapter 9, p. 162.)

corrected to Lyman's pH scale, gave $8.26 \cdot 10^{-10}$, that is, 1.6% higher. Skirrow, in his comprehensive 1975 review, does not explain why the GEOSECS results were inconsistent in such a systematic way.

Surface water variations are primarily controlled by the biological activity there: phytoplankton consume CO_2 and produce oxygen during daylight hours, and in the dark produce CO_2 by respiration; zooplankton feed on the phytoplankton and respire, producing CO_2 near the surface during the night when they migrate from deeper waters. Organic detritus is metabolized by microbes as well as by larger creatures, producing CO_2 throughout the water column, but primarily in the top 100–500 m. Carbon dioxide is absorbed from the atmosphere when phytoplankton photosynthesis depletes the CO_2 in surface waters below saturation.

Many planktonic species, both plant and animal, produce calcium carbonate skeletal structures, and in doing so they remove both alkalinity (Ca^{++}) and CO_2 from solution. When calcium carbonate detritus sinks to deeper waters it may redissolve (K_{s0} is larger at higher pressures, as discussed in the next section), but if it is buried in the sea floor sediments, it is effectively removed from the aquatic cycle. Carbonate and calcium ion are replenished by river input.*

---

* See Holland, H. D., op. cit. for a recent and comprehensive survey.

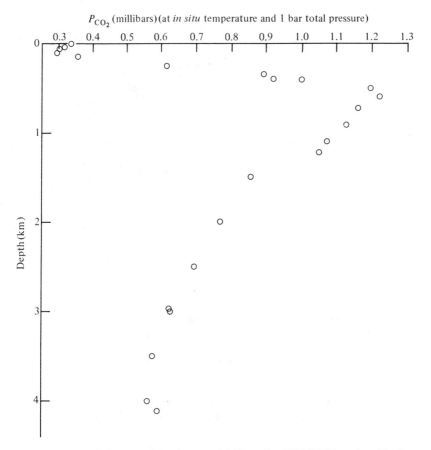

Fig. 5.5.   Equilibrium partial pressure of CO_2 at the GEOSECS station. The large increase in the first few hundred meters below the surface is a result of planktonic respiration. Compare with the small increase in alkalinity (Fig. 5.3) over the same depth range.

In Chapter 2, you saw that alkalinity is not affected by gain or loss of CO_2. Compare P_{CO_2} (Fig. 5.5), which increases by a factor of 4 from the surface to 500 m, while A_T increases by only 5% in the same depth range (Fig. 5.3). The increase in A_T (by 10% in deep water) is primarily the result of calcium carbonate dissolution, but a small contribution also results from release of inorganic silicate and phosphate ions as organic matter decomposes.

You will recall from Chapter 4 that loss or gain of $CaCO_3$ does not affect $D = 2(C_T - [Ca^{++}])$. Since alkalinity is the difference between the concentrations of cations and anions that do not participate in acid–base reactions over the pH range 4.5 to 8.5, a change in calcium ion concentration $\Delta[Ca^{++}]$ will be

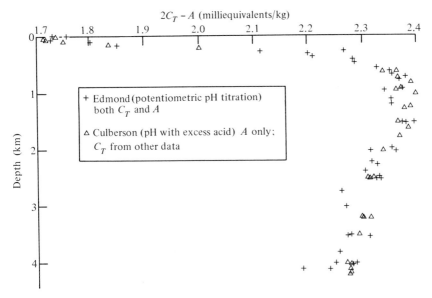

Fig. 5.6. The combination $2C_T - A$ as a function of depth at the GEOSECS station. Changes in this function (compare with D in Chapter 4) follow changes in CO_2 (Fig. 5.5) but are not affected by dissolution or precipitation of $CaCO_3$.

reflected in a change of alkalinity $\Delta A = 2\Delta[Ca^{++}]$ and consequently

$$\Delta D = \Delta(2C_T - A) + \cdots.$$

(5.36)

Figure 5.6 shows how $2C_T - A$ varies with depth. It follows closely the shape of the P_{CO_2} curve (Fig. 5.5). Unlike P_{CO_2}, however, it is not affected by $CaCO_3$ precipitation and dissolution.

THE LYSOCLINE

Calcium carbonate is produced by plankton in almost all oceanic surface waters, but in surface sediments from some deep parts of the ocean it is almost entirely absent. The upper part of the depth range over which the transition from carbonate-containing sediments to carbonate-free sediments occurs is called the *lysocline* (a steep gradient of solubility, by analogy with the thermocline). This transition is most clearly seen by comparing the species of foraminifera (calcareous protozoan plankton) found in the sediments. Those species that produce delicate, spiky shells are less frequent in deeper waters than those that are more robust and not so easily dissolved.

Deeper than the lysocline is the *calcite compensation depth*, where the downward transport of calcite detritus is presumably balanced by its dissolution in the

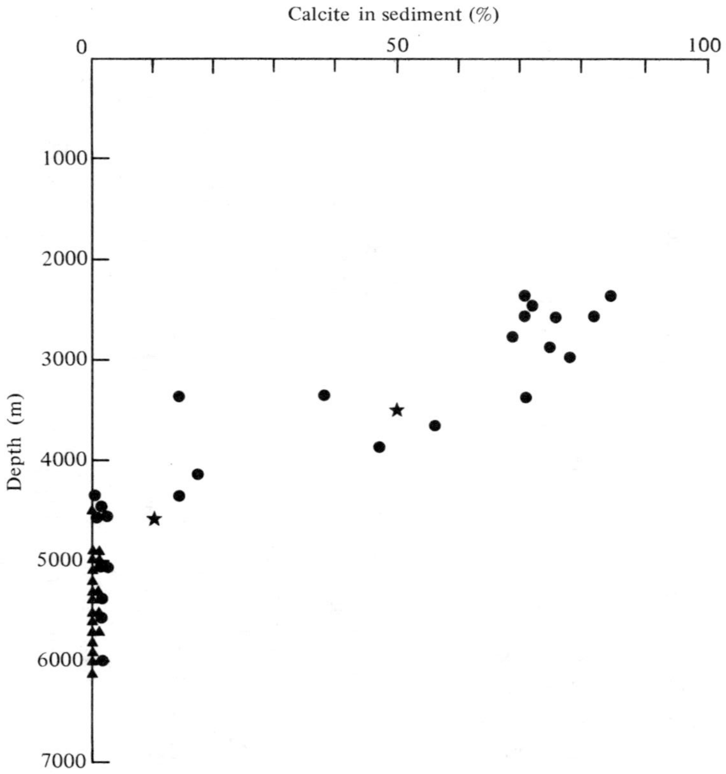

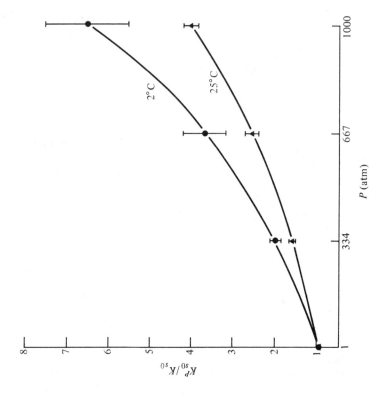

Fig. 5.7. Sediment composition versus depth for the central North Pacific Ocean (Area 4), showing the transition from substantial fractions of calcite to essentially no calcite below 4500 m, the calcite compensation depth. (From Berger, W. H., Adelseck, C. G., and Mayer, L. A. 1976. *J. Geophys. Res.* 81:2617–2627.)

Fig. 5.8. Pressure dependence of solubility product for calcite, expressed as the ratio of K_{s0} at pressure P to K_{s0} at 1 atm in seawater. Error bars are ± 2 standard deviations. (From Ingle, S. E. 1975. *Mar. Chem.* 3:301–319.)

undersaturated bottom waters. Little or no calcium carbonate is found in the sediments below this depth (Fig. 5.7). The lysocline and compensation depth occur at greater depths (5000–6000 m) in the Atlantic than in the Pacific (4000–5000 m), and at greater depths in equatorial than in polar or nearshore regions.

Deep waters are undersaturated with calcium carbonate primarily because its solubility increases strongly with increasing pressure (Fig. 5.8) and salinity (Fig. 5.9) and slightly with decreasing temperature (Fig. 5.9). This decrease of

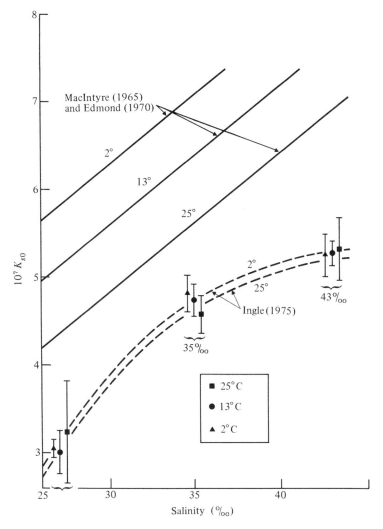

Fig. 5.9. Salinity and temperature dependence of the solubility product for calcite in seawater. Error bars are ±2 standard deviations.

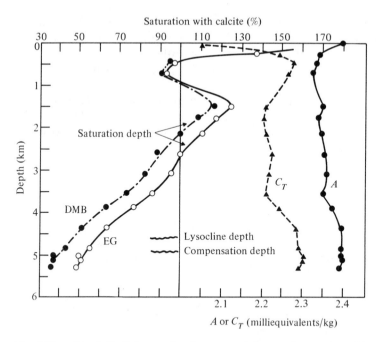

Fig. 5.10.   Total alkalinity, total carbonate, and degree of saturation with calcite as a function of depth. Data obtained at station Circe 245, (7°9′S, 21°21′W) in the equatorial Atlantic; EG stands for Edmond, J. M., and Gieskes, J. M. T. M., 1970. Geochim. Cosmochim. Acta 34:1261–1291; DMB stands for Gieskes' recalculation. (From Gieskes, J. M. 1975. In *The Sea*, E. D. Goldberg, ed., vol. 5. New York: Wiley, Chapter 2, p. 141.)

saturation with depth is not due to a change in total carbonate or alkalinity. At the Equatorial Atlantic station where the data illustrated in Fig. 5.10 were taken, neither of these solution parameters varies much with depth, whereas the degree of saturation with calcite varies from more than 40% supersaturated at the surface to about 40 or 50% undersaturated at the bottom at 5300 m. The two saturation curves illustrate some of the uncertainty inherent in these thermodynamic data, but the trend is not substantially affected.

From Fig. 5.10 you can also see the depth (at 2000 to 2500 m) at which seawater is calculated to be just saturated with calcite. Some uncertainty results from the theory, but this depth is significantly shallower than the observed lysocline (obtained by examination of forams in sediments) or compensation depth (obtained from the calcite content of sediments, as in Fig. 5.7). This difference between calculated and observed depths reflects the relatively slow rate at which calcite particles dissolve as they fall through undersaturated seawater. Some investigators* have suggested that at a critical degree of undersaturation (about 50 to 70%) there is a change in mechanism that allows much more rapid dissolution, and this change in mechanism is responsible for the lysocline.

EXAMPLE 9   Show that surface seawater is supersaturated with calcite and deep waters are undersaturated. Some typical data for surface and deep ocean waters are given in the table.

Parameter	Surface	Deep
Depth	0 m	6730 m
Pressure	1 atm	667 atm
Temperature	25°C	2°C
Salinity	35‰	35‰
$[Ca^{++}]$	$1.03 \cdot 10^{-2}$	$1.03 \cdot 10^{-2}$ mole/kg
$A_c (= 0.964 A_T)$	$2.17 \cdot 10^{-3}$	$2.36 \cdot 10^{-3}$ mole/kg
C_T	$1.95 \cdot 10^{-3}$	$2.35 \cdot 10^{-3}$ mole/kg
K_{s0} (calcite)	$10^{-6.34}$	$10^{-5.75}$
K_1'	$10^{-6.00}$	$10^{-6.49}$
K_2'	$10^{-9.12}$	$10^{-9.64}$

Note that the ratio of calcium ion to salinity is constant to within 1% throughout the oceans, and since salinity changes by about 1‰ over 6000 m, both have been taken as constant. Alkalinity is taken from Fig. 5.3 and C_T from Fig. 5.4. The solubility products at atmospheric pressure were obtained from Fig. 5.9 and corrected to 667 atm with Fig. 5.8. The values of K_1' and K_2' at atmospheric pressure are from the data of Mehrbach, corrected to 667 atm with the pressure coefficients of C. Culberson and R. M. Pytkowicz, 1968. (*Limnol. Oceanogr.* 13:403–417; see also Skirrow, op. cit., pp. 66–68 and Appendixes A9.9 and A9.10). Pressure and depth (z) are related by the density of seawater:

$$P = \frac{(1.0235 \text{ g/cm}^3)(10^{-3} \text{ kg/g})(10^2 \text{ cm/m})}{(1.03323 \text{ kg/atm} \cdot \text{cm}^2)} \cdot z = (9.906 \cdot 10^{-2} \text{ atm/m})z.$$

---

* Morse, J. W. and Berner, R. A. 1972. *Am. J. Sci.* 272:840–851; Morse, J. W. and Berner, R. A. 1979. In *Chemical Modeling in Aqueous Systems*, E. A. Jenne, ed. American Chemical Society Symposium Series No. 93, Chapter 24. Washington DC: American Chemical Society.

To test the degree of saturation, you have to calculate $[CO_3^=]$ from A and C_T. This is best done by using the relations from Chapter 2. Combine (2.16), (2.18), (2.22), (2.23), (2.39), and (2.42), summarized on pp. 228–229, to get

$$A_c - C_T = [CO_3^=] - [CO_2] = \frac{C_T(K_1'K_2' - 10^{-2pH})}{10^{-2pH} + K_1'10^{-pH} + K_1'K_2'}. \qquad (5.37)$$

This equation can be rearranged into the quadratic form

$$A_c 10^{-2pH} + (A_c - C_T)K_1'10^{-pH} - (2C_T - A_c)K_1'K_2' = 0. \qquad (5.38)$$

Since pH is near 8, and both A_c and C_T are of the order of 10^{-3}, the first term is almost negligible for the surface but not for the deep-sea conditions. Substituting the values given at the beginning of the example, you will find that pH = 8.25 for the surface waters and 8.10 for the deep waters. Using these values in the expression for carbonate ion concentration obtained from (2.29), (2.39), and (2.42)

$$[CO_3^=] = \frac{C_T K_1'K_2'}{10^{-2pH} + K_1'10^{-pH} + K_1'K_2'} \qquad (5.39)$$

you can obtain the following results:

Parameter	Surface	Deep
pH	8.25	8.10
$[CO_3^=]$	$10^{-3.63}$	$10^{-4.18}$
$[Ca^{++}][CO_3^=]$	$10^{-5.62}$	$10^{-6.17}$
$[Ca^{++}][CO_3^=]/K_{s0}$	$10^{+0.72}$	$10^{-0.42}$

The last parameter is a measure of the degree of supersaturation, the ratio of the actual concentration product to the equilibrium solubility product K_{s0}. This result is in qualitative agreement with observations such as in Fig. 5.10: the surface waters are supersaturated, while the deep waters are undersaturated. ▪

Aragonite is slightly more soluble than calcite (Table 5.5), and so it is not surprising that pteropod (planktonic mollusc) shells made of aragonite are only found at depths considerably more shallow than the lysocline or calcite compensation depth (2000–3000 m in the Atlantic, 500–1500 m in the Pacific). Nevertheless, Berner[*] has suggested that the sedimentation and dissolution of these

---

[*] Berner, R. A. 1977. Sedimentation and dissolution of pteropods in the ocean. In *The Fate of Fossil Fuel CO_2 in the Oceans*, N. R. Andersen and A. Malahoff, eds. New York: Plenum, pp. 505–542.

aragonite shells constitute as much as 50% of the flux of calcium carbonate from surface waters to relatively deep waters.

A further complication in studies of the precipitation and dissolution of calcium carbonate in the oceans is the presence of magnesium in seawater at concentrations five times as large as calcium (see Table 5.4). Many calcium carbonate sediments and limestone rocks contain several percent of magnesium substituting for calcium in the lattice (magnesian calcites), and a relatively common mineral in ancient rocks is dolomite $MgCa(CO_3)_2$.

The effects of magnesium on calcite equilibria and solution kinetics are complicated. For example, a recent study* of the behavior of calcite in surface seawater (four times supersaturated with calcite) shows some of this complexity. When pure calcite is first introduced to continually renewed seawater, a layer of carbonate grows on its surface, but this carbonate is not pure calcite; it has a magnesium content (up to 10 mole % replacing Ca) that increases with the square root of the thickness of the layer. At the same time, the solubility of magnesian calcite increases with its magnesium content. If this process continued, eventually the carbonate deposited would be precisely saturated with respect to seawater, and growth would stop.

However, the carbonate deposit continues to grow, because the inner layers recrystallize so as to exclude magnesium and form crystals that approach pure calcite in composition. Under some conditions, carbonate deposits with higher magnesium contents (about 8%) may recrystallize to form aragonite. One of the main factors controlling the rate of the recrystallization process is grain size. Finely powdered pure calcite, recently reacted with seawater, recrystallizes on a time scale of hours to days; but the complete conversion to calcite of biologically produced magnesian calcites or aragonite in large complex structures may take millions of years.[†]

EFFECT OF INCREASED ATMOSPHERIC CO$_2$ ON THE OCEANS

Burning of coal and other human activities have caused a significant increase in atmospheric carbon dioxide (Fig. 5.11), and this increase is expected to continue for at least another century before other energy sources are fully developed. Most of this additional carbon dioxide will find its way into the water and sediments of the deep oceans.

The most rapid process by which the oceans remove CO_2 from the atmosphere is homogeneous dissolution and reaction with $CO_3^=$ to produce HCO_3^-. These

---

* Wollast, R., Garrels, R. M., and Mackenzie, F. T., 1980. *Am. J. Sci.* 280:831–848.

[†] I refer you to the geochemical literature for further discussion of these topics, especially Bathurst (op. cit.), and Skirrow (op. cit.).

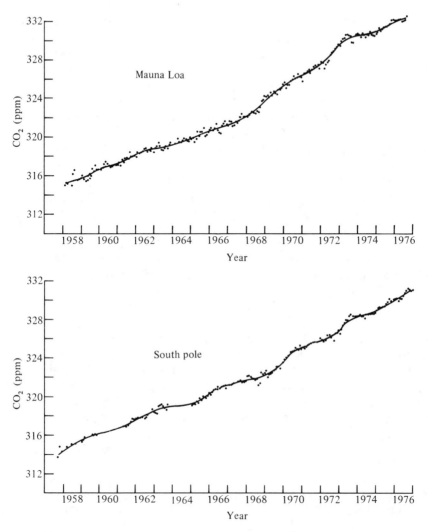

Fig. 5.11.   Seasonally adjusted concentration of atmospheric CO_2 at Mauna Loa Observatory, Hawaii, and at the South Pole. Points indicate seasonally adjusted monthly averages. Curves are smooth piecewise fits to the points. These data show clearly the global increase of almost 20 ppm between 1958 and 1976. (From Bacastow, R. B. and Keeling, C. D. 1979. *Models to Predict Atmospheric CO_2*. Department of Energy Report CONF-770385, p. 75.)

chemical reactions proceed in the laboratory as fast as solutions can be mixed; the upper 100 m mixed layer of the ocean can reach equilibrium with the atmosphere in one year. The calculations below will use the equilibria you are already familiar with to evaluate this process quantitatively.

Additional processes for CO$_2$ removal include increased weathering of carbonate and silicate rocks, dissolution of CaCO$_3$ particles and sediments in the oceans, and possibly increased photosynthesis in areas where nutrients are in excess. These processes are shown schematically in Fig. 5.12. The rates and extent of these more complicated processes, as well as the effect of increased CO$_2$ on climate, are beyond the scope of this book, and I refer you to the recent literature for details.*

In discussions of how much CO$_2$ can be absorbed by homogeneous reaction with the mixed layer of seawater, the Revelle buffer factor[†]

$$B = \frac{C_T}{P}\left(\frac{\partial P}{\partial C_T}\right)_A = \left(\frac{\partial \log P}{\partial \log C_T}\right)_A \tag{5.40}$$

is often used. The derivative of partial pressure P of CO$_2$ with respect to total carbonate in solution C_T is taken at constant alkalinity A to reflect the fact that open ocean surface waters are mostly supersaturated with CaCO$_3$ and have few sources of alkalinity that might react rapidly with CO$_2$ and provide additional buffering capacity. Therefore B, as defined above, represents a lower limit.

The basic equations come from Chapter 2: combine (2.1), (2.2), and (2.3) with (2.18) to get

$$C_T = [CO_2] + [HCO_3^-] + [CO_3^=] = K_H P\left(1 + \frac{K_{a1}}{[H^+]} + \frac{K_{a1}K_{a2}}{[H^+]^2}\right) \tag{5.41}$$

and with (2.16) to get

$$A = [HCO_3^-] + 2[CO_3^=] + [OH^-] - [H^+]$$
$$= K_H P\left(\frac{K_{a1}}{[H^+]} + \frac{2K_{a1}K_{a2}}{[H^+]^2}\right) + \frac{K_w}{[H^+]} - [H^+]. \tag{5.42}$$

---

* Broecker, W. S., Li, Y.-H., and Peng, T. H. 1971. In *Impingement of Man on the Oceans*, D. W. Hood, ed. New York: Wiley, pp. 287–324; Keeling, C. D. 1973. In *Chemistry of the Lower Atmosphere*, S. I. Rasool, ed. New York: Plenum, pp. 251–329; Andersen, N. R. and Malahoff, A., eds. 1977. *The Fate of Fossil Fuel CO$_2$ in the Oceans*. New York: Plenum; Elliott, W. P. and Machta, L., eds. 1979. *Workshop on the Global Effects of Carbon Dioxide from Fossil Fuels*. US Dept. of Energy CONF-770385; Holland, H. D., op. cit., Chapter 6.

[†] Sundquist, E. T., Plummer, L. N., and Wigley, T. M. L. (1979. *Science* 204:1203–1205) provide the basis for this discussion. They use some different notation: $B = B_{Hom}$ (homogeneous buffer capacity), $C_T = TCO_2$, $P = pCO_2$, $A = TA$. They also include terms for the contribution of borate to the total alkalinity (which I have left out for simplicity) to obtain Eq. (5.49) given on p. 146.

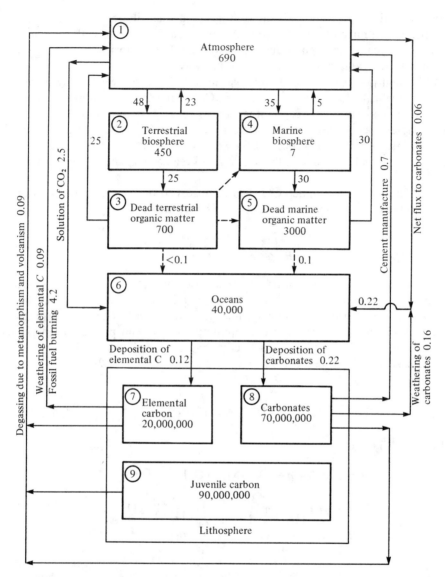

Fig. 5.12. The global carbon cycle expressed as a box model. The contents of the boxes are in units of 10^{15} g of carbon and transfer rates are contemporary estimates in units of 10^{15} g of carbon per year. Note especially that the combination of fossil-fuel burning (4.2) and cement manufacture (0.7) greatly exceeds the atmospheric input due to metamorphism, vulcanism, weathering of elemental C, and is about twice the rate at which CO_2 is absorbed by the oceans (2.5). The terrestrial and marine biospheres transfer more carbon to and from the atmosphere, but are essentially in balance. (From Holland, H. D. 1978. *Chemistry of the Atmospheres and Oceans.* New York: Wiley, Chapter 6.)

You can solve Eq. (5.42) for P in terms of A and $[H^+]$, substitute this expression in (5.41) to obtain C_T as a function of A and $[H^+]$, and from these two equations compute B as defined in (5.40) from

$$B = \frac{C_T(\partial P/\partial[H^+])_A}{P(\partial C_T/\partial[H^+])_A}. \tag{5.43}$$

Differentiate (5.41) with respect to $[H^+]$, holding A constant

$$\left(\frac{\partial C_T}{\partial[H^+]}\right)_A = K_H P\left(-\frac{K_{a1}}{[H^+]^2} - \frac{2K_{a1}K_{a2}}{[H^+]^3}\right) + K_H\left(1 + \frac{K_{a1}}{[H^+]} + \frac{K_{a1}K_{a2}}{[H^+]^2}\right)\left(\frac{\partial P}{\partial[H^+]}\right)_A,$$

and substitute back for the combinations of K and $[H^+]$ in terms of the concentration. This gives the denominator of (5.43) and involves no approximations yet:

$$\left(\frac{\partial C_T}{\partial[H^+]}\right)_A = -\frac{1}{[H^+]}([HCO_3^-] + 2[CO_3^=])$$

$$+ \frac{1}{P}([CO_2] + [HCO_3^-] + [CO_3^=])\left(\frac{\partial P}{\partial[H^+]}\right)_A. \tag{5.44}$$

To get the term $(\partial P/\partial[H^+])$, required for the last part of (5.44) and the numerator of (5.43), take the derivative of (5.42). Since A is constant,

$$\left(\frac{\partial A}{\partial[H^+]}\right)_A = 0 = K_H P\left(-\frac{K_{a1}}{[H^+]^2} - \frac{4K_{a1}K_{a2}}{[H^+]^3}\right) - \frac{K_w}{[H^+]^2}$$

$$- 1 + K_H\left(\frac{K_{a1}}{[H^+]} + \frac{2K_{a1}K_{a2}}{[H^+]^2}\right)\left(\frac{\partial P}{\partial[H^+]}\right)_A.$$

Identify the products as concentrations:

$$\frac{1}{P}([HCO_3^-] + 2[CO_3^=])\left(\frac{\partial P}{\partial[H^+]}\right)_A$$

$$= \frac{1}{[H^+]}([HCO_3^-] + 4[CO_3^=] + [OH^-] + [H^+])$$

or

$$\left(\frac{\partial P}{\partial[H^+]}\right)_A = \frac{P}{[H^+]}\frac{[HCO_3^-] + 4[CO_3^=] + [OH^-] + [H^+]}{[HCO_3^-] + 2[CO_3^=]}. \tag{5.45}$$

When the full expression (5.45) for the derivative of CO$_2$ partial pressure with respect to $[H^+]$ is substituted in (5.44), you get

$$\left(\frac{\partial C_T}{\partial[H^+]}\right)_A = \frac{1}{[H^+]}\Bigg(-[HCO_3^-] - 2[CO_3^=]$$

$$+ C_T\frac{[HCO_3^-] + 4[CO_3^=] + [OH^-] + [H^+]}{[HCO_3^-] + 2[CO_3^=]}\Bigg).$$

Multiplying out this expression, setting $C_T = [CO_2] + [HCO_3^-] + [CO_3^=]$, you will note that *all* negative terms are cancelled by similar positive terms:

$$\left(\frac{\partial C_T}{\partial [H^+]}\right)_A = \frac{[HCO_3^-]([CO_2]+[CO_3^=])+4[CO_2][CO_3^=]+C_T([H^+]+[OH^-])}{[H^+]([HCO_3^-]+2[CO_3^=])}. \tag{5.46}$$

By taking only the largest terms in (5.46), you can eliminate all but the first two terms of the numerator and the first of the denominator; $[HCO_3^-]$ cancels to give

$$\left(\frac{\partial C_T}{\partial [H^+]}\right)_A = \frac{[CO_2]+[CO_3^=]}{[H^+]} + \cdots, \tag{5.47}$$

and when combined with (5.45) (similar approximations hold), the final result is

$$B = \frac{C_T}{P}\frac{P}{[H^+]}\frac{[H^+]}{[CO_2]+[CO_3^=]} = \frac{C_T}{[CO_2]+[CO_3^=]} + \cdots. \tag{5.48}$$

EXAMPLE 10   For a typical seawater, pH $= 8$, $A \cong [HCO_3^-] = 10^{-2.7}$, $C_T = 10^{-2.7}$, $K_1' = 10^{-6.0}$, $K_2' = 10^{-9.1}$, find B. From (2.39) and (2.42) you get:

$$[CO_2] = \frac{A(10^{-pH})}{K_1'} = 10^{-4.7}, \qquad [CO_3^=] = \frac{AK_2'}{10^{-pH}} = 10^{-3.8},$$

and hence $B = 11.2$. Values obtained in the field at $25°C$ are approximately 9.5 (see Fig. 5.13).   ■

Including the contribution of borate to the buffer capacity makes a significant difference. The expression corresponding to (5.48) is

$$B = C_T\left([CO_2] + [CO_3^=] + \frac{\gamma[HCO_3^-] - 4[CO_3^=]^2}{[HCO_3^-] + 4[CO_3^=] + \gamma}\right)^{-1}, \tag{5.49}$$

where

$$\gamma = \frac{C_B K_B' 10^{-pH}}{(K_B' + 10^{-pH})^2}.$$

Here C_B is the borate concentration of seawater, and K_B' the acidity constant for borate (see Eqs. 3.51 to 3.53, pp. 63–64, and Table 5.5)

In Fig. 5.13, the calculated curve of Eq. (5.49) is compared with the GEOSECS field data from both Pacific and Atlantic Oceans.[*] Agreement is excellent within the uncertainties introduced by natural variations and by errors in the equilibrium constants.

---

[*] Sundquist et al., op. cit.

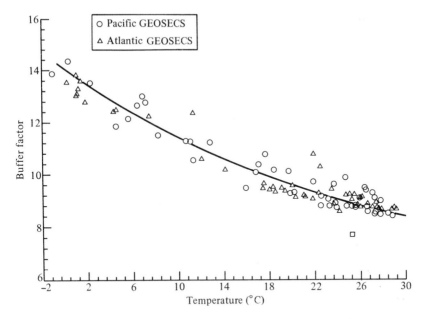

Fig. 5.13. Homogeneous buffer factor (Eq. (5.40)) as a function of temperature. (From Sundquist, Plummer, and Wigley. 1979. *Science* 204:1203–1205.)

A value of about 10 for B tells you that if surface alkalinity remains constant, a change of 10% in the atmospheric partial pressure of CO_2 is required to produce a 1% change in the total CO_2 content of seawater in equilibrium with that atmosphere. Earlier, I pointed out that, although the time scale for mixing of the deep ocean with its surface waters is one or two thousand years, the top few hundred meters (mixed layer) mixes on a time scale of less than a year. How much of the recent increase in atmospheric CO_2 might have been taken up by this mixed layer? Broecker et al.* estimated the preindustrial atmosphere (290 ppm CO_2) to have approximately 145 moles of CO_2 above each square meter of ocean surface, and the mixed layer of the ocean to contain about 900 mole/m^2 of total carbonate. Thus a 10% increase in atmospheric CO_2 (by 30 ppm or 15 mole/m^2) would produce an increase of about 1% in total dissolved carbonate (9 mole/m^2). By this mechanism, therefore, the oceans can absorb, within a year or two, about half (9 out of 15) the increase in atmospheric CO_2. Attempts to measure the predicted lag (the oceans should have an equilibrium P_{CO_2} that is 1 to 3 ppm less than the atmosphere) have unfortunately been obscured by natural variations in these quantities.

It is interesting to note that an increase in the CO_2 content of seawater can enhance its capacity for moderating atmospheric CO_2. The homogeneous buffer factor given by Eq. (5.48) or (5.49) increases with increasing C_T to a maximum

---

* Broecker, W. S., Takahashi, T., Simpson, H. J., and Peng, T. H. 1979. *Science* 206:409–418.

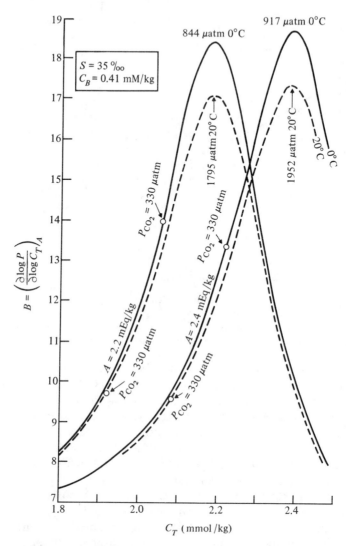

Fig. 5.14. Homogeneous buffer factor as a function of total dissolved carbonate and alkalinity. Note that the current open-ocean conditions (lower left circle at 20°C, $P = 330$ μatm, and $A = 2.2$ mEq/kg) produce a much lower buffer factor than that achieved if additional CO_2 is absorbed by the oceans. (From Takahashi, T. 1979. *Carbon Dioxide Chemistry in Ocean Water.* Dept. of Energy Report CONF-770385, p. 63.)

of approximately 18, twice the value for the present oceans. Figure 5.14 shows this dependence of B on C_T for temperatures of $0°C$ and $20°C$, and alkalinity of either 2.2 or 2.4 mEq/kg.* These ranges account for nearly all open ocean water. Four circles show the values corresponding to present atmospheric CO_2: 330 μatm or 330 ppm. Increasing C_T at constant A and T increases B; increasing A at constant C_T and T decreases B. The maximum buffer capacity is achieved at 2.5 to 6 times the present P_{CO_2}, depending on temperature and alkalinity.

On a longer time scale, the condition of constant alkalinity used in the above derivation and in the two-box model of Problem 50 will not apply, because carbon dioxide will react with calcium carbonate and other minerals. This might be most easily seen as increased weathering of continental limestone and the reaction of CO_2 with shallow-water carbonate sediments soluble enough to come to equilibrium (see the discussion of magnesian calcites, p. 141, and Problem 49.) Eventually the excess CO_2 would be transferred to deeper waters and would react with the surface sediments there. Atmospheric CO_2 will also react (less rapidly) with other minerals such as Ca and Mg silicates.

One possible consequence is that the lysocline or calcite compensation depth might migrate to more shallow waters as the deeper waters became less saturated with $CaCO_3$. Another possible consequence is that the production of calcite and aragonite by planktonic organisms in near-surface waters might be diminished as the water becomes less supersaturated. However, the total capacity of rock weathering and the ocean, including sediments, for absorbing CO_2 in the long term (over 1000 years) far exceeds any reasonable estimate for the amount of excess CO_2 to be introduced by combustion of fossil fuels in the next two or three centuries.

While this last point about the buffering capacity of rocks and oceanic sediments may be encouraging, it does not eliminate the possibility that short-term changes in the atmospheric CO_2 might create significant changes in planktonic communities as well as in the global climate.

FRESH WATERS AND ESTUARINE WATERS

Like river and ocean waters, fresh waters (and estuarine waters, which are dynamic mixtures of fresh and saline waters) derive much of their carbonate content from the atmosphere, from the weathering of rocks, and from microbial decomposition of organic matter. They provide CO_2 to phytoplankton and macro-algae for photosynthetic growth. One difference between estuarine and ocean water is the former's much higher concentration of sediments, derived from rivers.

Another important factor is the salinity of the water, which you know has a large effect on the acid–base equilibria of the carbon dioxide system. The simplest

---

* Takahashi, T. 1979. In *Workshop on the Global Effects of CO₂ from Fossil Fuels,* W. P. Elliott and L. Machta, eds. U.S. Dept. of Energy CONF-770385, pp. 63–71.

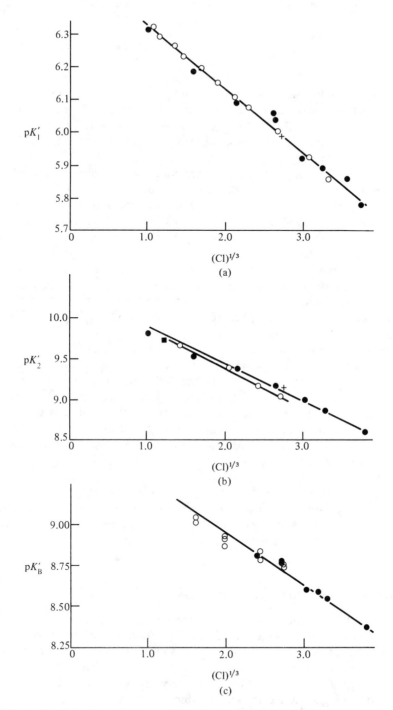

Fig. 5.15. Chlorinity dependence at 20°C of the carbonate equilibrium constants and the borate constant. Data from Lyman (op. cit.) (●), Disteche, A., and Disteche, S., 1967. J. Electrochem. Soc. 114:330–340 (+), and Buch, K., Harvey, H. W., Wattenberg, H. and Gripenberg, S. 1932. Rapp. Cons. Explor. Mer. 79:1–70; also Buch, K. 1938. Acta Acad. abo. 11: No. 5. (○), adjusted to make them compatible with those of Lyman. (From Skirrow, G. 1975. In *Chemical Oceanography*, vol. 2. New York: Academic Press, Chapter 9, pp. 47–48 [after Edmond and Gieskes, op. cit.].)

chemical model follows from the assumptions that river water and ocean water mix without losing or gaining any alkalinity or total carbonate.*

The equilibrium constants, as you have seen, vary with salinity and temperature. A set of empirical equations (Skirrow, op. cit.) approximately describes this dependence:

$$-\log K'_1 = 3404.71/T + 0.032786T - 14.7122 - 0.19178\,(Cl)^{1/3},$$
$$-\log K'_2 = 2902.39/T + 0.02379T - 6.4710 - 0.4693\,(Cl)^{1/3}, \qquad (5.50)$$
$$-\log K'_B = 2291.90/T + 0.01756T - 3.3850 - 0.32051\,(Cl)^{1/3},$$

where T is temperature in $°K$, and Cl is chlorinity expressed in $‰$ (see Fig. 5.15).

Let b be the fraction of the water that originated in the ocean. This quantity is obtained experimentally by measuring the chlorinity:

$$b = \frac{Cl - Cl^r}{Cl^o - Cl^r}. \qquad (5.51)$$

Here Cl^r is the chlorinity of the river water and Cl^o is the chlorinity of the ocean water.

Then C_T and A for the mixed estuarine water are linear combinations of values for ocean and river water:

$$C_T = bC_T^o + (1 - b)C_T^r \qquad (5.52)$$

$$A = bA^o + (1 - b)A^r. \qquad (5.53)$$

Here A can represent either total alkalinity A_T or carbonate alkalinity A_c. Although A_T is the usual measurement, A_c is required for the subsequent calculations, and is obtained by correcting for boron (see pp. 52 and 63):

$$A_c = A_T - \frac{K'_B C_B}{K'_B + 10^{-\text{pH}}} + \cdots, \qquad (5.54)$$

where $C_B \cong 2.2 \cdot 10^{-2}\,Cl$ and K'_B is given by the empirical Eq. (5.50) above.

Thus measurements of total alkalinity, chlorinity, and pH in the river and ocean waters should allow prediction of alkalinity, pH, and C_T in any mixed estuarine water if only its chlorinity is measured. Note that for any such mixed water, a value of Cl gives a value for b by means of (5.51). In turn, b gives values for C_T and A_c by (5.52) and (5.53). Their ratio is

$$Q = \frac{C_T}{A_c} = \frac{bC_T^o + (1 - b)C_T^r}{bA_c^o + (1 - b)A_c^r}. \qquad (5.55)$$

But Q can also be calculated knowing only pH and the equilibrium constants (Eqs. (2.29) and (3.11) pp. 26 and 43):

$$Q = \frac{C_T}{A_c} = \frac{[CO_2] + [HCO_3^-] + [CO_3^=]}{[HCO_3^-] + 2[CO_3^=]} = \frac{10^{-2\text{pH}} + 10^{-\text{pH}}K'_1 + K'_1 K'_2}{10^{-\text{pH}}K'_1 + 2K'_1 K'_2}. \qquad (5.56)$$

---

* Mook, W. G. and Koene, B. K. 1975. *Estuarine Coastal Mar. Sci.* 3:325–336.

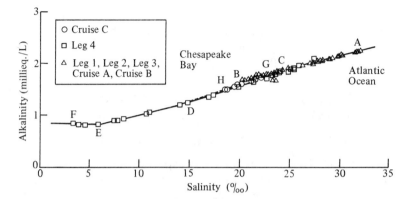

Fig. 5.16. Relationship between alkalinity and salinity in the southern Chesapeake Bay and James River estuary. Line *AB* represents mixing between inflowing Atlantic Ocean water (*A*) and water from northern Chesapeake Bay (*B*). Line *CDEF* represents mixing of James River water with Chesapeake Bay water. The discontinuities in slope correspond to the mixing of tributaries with the main estuarine water. (From Wong, G. T. F. 1979. *Limnol. Oceanogr.* 24: 970–77.)

Therefore, once Q is known from (5.55), and values for K_1' and K_2' are determined from chlorinity in (5.50), pH is obtained from the quadratic (rearrangement of (5.56) above):

$$10^{-2\mathrm{pH}} + 10^{-\mathrm{pH}}K_1'(1 - Q) + K_1'K_2'(1 - 2Q) = 0. \tag{5.57}$$

Figure 5.16 shows that alkalinity varies essentially linearly over a wide range of salinity.* pH varies nonlinearly but is constant at 8.0–8.2 above 15‰ salinity.

The same principles apply to the mixing of ocean water masses, lakes, and groundwaters. Of course, if substantial $CaCO_3$ sediment is present, A and C_T will vary as this sediment dissolves, precipitates, or settles out, and the models must be made more complicated.

Predicting whether a mixture of two solutions will be saturated with $CaCO_3$ is an important problem in geochemistry, but is complicated by nonlinear relationships between most of the important factors. The easiest of these to demonstrate involves the solubility product. Assume that to a first approximation (see below) both $[Ca^{++}]$ and $[CO_3^=]$ can be linearly interpolated:

$$[Ca^{++}] = y[Ca^{++}]_1 + (1 - y)[Ca^{++}]_2 + \cdots, \tag{5.58}$$

$$[CO_3^=] = y[CO_3^=]_1 + (1 - y)[CO_3^=]_2 + \cdots, \tag{5.59}$$

where y is the volume fraction of the first end-member solution; the other end-member is subscripted 2. If both the end-member solutions are saturated with

---

* Wong, G. T. F. 1979. *Limnol. Oceanog.* 24:970–77.

calcium carbonate,

$$[Ca^{++}]_1[CO_3^=]_1 = [Ca^{++}]_2[CO_3^=]_2 = K_{s0}. \tag{5.60}$$

Again, this is an approximation, since the ionic strength and composition of the two solutions is not necessarily identical, and so K_{s0} is not exactly the same for the two solutions.

Substituting $[CO_3^=]$ from (5.60) in (5.59) and combining the result with (5.58), the degree of saturation can be obtained:

$$\Omega = \frac{[Ca^{++}][CO_3^=]}{K_{s0}} = 1 + (R - 2)y - (R - 2)y^2,$$

where

$$R = \frac{[Ca^{++}]_2}{[Ca^{++}]_1} + \frac{[Ca^{++}]_1}{[Ca^{++}]_2}. \tag{5.61}$$

Note that $\Omega = 1.0$ for $y = 0$ and $y = 1$ (the two end members) and reaches a maximum when $y = 1/2$:

$$\Omega = \frac{R + 2}{4} \geq 1. \tag{5.62}$$

If the first and second end-member solutions are identical, $R = 2$, $\Omega = 1$, and the mixed solution remains saturated. If they differ in composition, R is greater than 2 (as you can verify with a few numerical examples) and $\Omega > 1$. Thus the mixture of two saturated solutions of different composition will normally yield a super-saturated solution.

As I mentioned above, there are other nonlinear effects to be considered. Mixing of two solutions containing calcium ion will not be perfectly linear, as I assumed in Eq. (5.58), because the degree of ion-pairing will in general be different in the mixture than in either of the end-member solutions. The degree of protona-tion and ion-pairing to which the carbonate ion is subject will change even more nonlinearly as the solutions are mixed.

Other related considerations are (a) the redistribution of carbonate species that occurs when waters of different pH, C_T, or P_{CO_2} are mixed; (b) the nonlinear shift of activity coefficients and redistribution of ion pairs (which can change ionic strength as well as pH, $[Ca^{++}]$, and $[CO_3^=]$) when solutions of different ionic composition are mixed; and (c) the nonlinear variation of equilibrium constants when solutions of different temperature are mixed.

Wigley and Plummer* have presented an algorithm for these calculations that assumes linear mixing for alkalinity and C_T to be the basic constraints, and uses an ion-pairing model to obtain the concentrations of other species. Initially, pH is obtained by linear interpolation, but on subsequent iteration it is calculated from the ion-pairing model. Their program WATMIX is said to converge after only three or four iterations.

---

* Wigley, T. M. L., and Plummer, L. N., 1976. *Geochim. Cosmochim. Acta* 40:989–995.

Table 5.8
Compositions of end-member solutions used in mixing calculations for Figs. 5.17 and 5.18*

Solution	t (°C)	pH	Alkalinity (mEq/kg)	Ca	Mg	Na	K (mmole/kg)	SO$_4$	Cl	P_{CO_2} (atm)	I (mole/kg)
1	10	7.335	4.332	2.166						10^{-2}	0.0065
2	10	7.990	1.928	0.964						10^{-3}	0.0029
3	25	7.313	3.398	1.699						10^{-2}	0.0051
4	10	7.071	2.572	14.639				13.353		10^{-2}	0.0404
5	10	7.448	5.762	1.341	1.540					10^{-2}	0.0085
6	10	7.386	5.560	2.780		59.762			59.762	10^{-2}	0.0683
Seawater	10	8.150	2.33	10.30	53.21	467.73	9.90	28.25	545.82	$10^{-3.3}$	0.6636

Solutions 1–6 are saturated with calcite; seawater is supersaturated with calcite ($[Ca^{++}][CO_3^=]/K_{s0} = 10^{+0.49}$).
solution 4 is saturated with gypsum; seawater is undersaturated ($[Ca^{++}][SO_4^=]/K_{s0}^G = 10^{-0.63}$);
solution 5 is saturated with dolomite; seawater is supersaturated ($[Ca^{++}][Mg^{++}][CO_3^=]^2/K_{s0}^D = 10^{+1.71}$).
* Wigley, T. M. L. and Plummer, L. N. op. cit. 995.

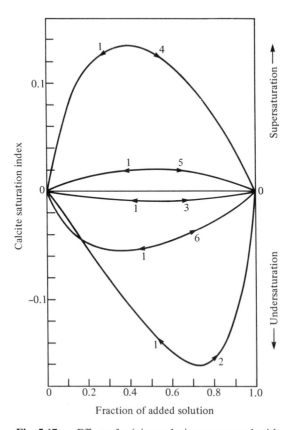

Fig. 5.17.   Effect of mixing solutions saturated with calcite (described in Table 5.9) on the degree of saturation with calcite, expressed as the index log $([Ca^{++}][CO_3^=]/K_{s0})$. Addition to solution 1 (calcium bicarbonate) of solution 2 (deficient in Ca) produces undersaturation, as do solution 3 (warmer, more dilute calcium bicarbonate) and solution 6 (calcium bicarbonate with added NaCl). Addition of solution 5 (saturated with dolomite as well as calcite) or solution 4 (saturated with gypsum as well as calcite) produces oversaturation. (From Wigley, T. M. L. and Plummer, L. N. 1976. *Geochim. Cosmochim. Acta* 40:989–995.)

Some results of Wigley and Plummer's calculations are presented in Table 5.8, Fig. 5.17, and Fig. 5.18. For example, Fig. 5.17 shows how two solutions, each saturated with calcite, can yield mixtures that are supersaturated (1 and 4, 1 and 5) or undersaturated (1 and 2, 1 and 3, 1 and 6). More complicated systems can yield both undersaturation and supersaturation in the mixtures. Figure 5.18 shows the degree of saturation with calcite when seawater is added to a solution obtained

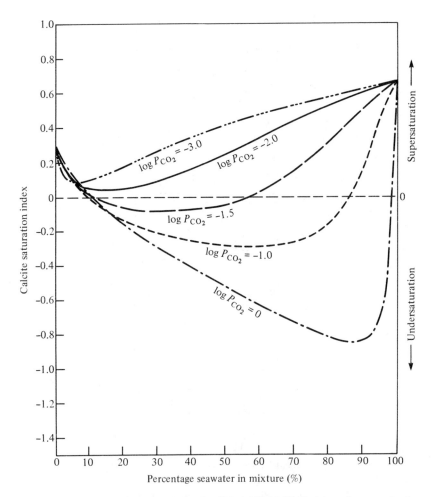

Fig. 5.18.   Calcite saturation index $\log([Ca^{++}][CO_3^{=}]/K_{s0})$ in mixtures of solutions twofold saturated with calcite with seawater (pH = 8.15) at 25°C. If P_{CO_2} is low, the mixed solutions remain supersaturated; but for $P_{CO_2} > 10^{-1.5}$, undersaturation occurs over at least part of the mixing range. Factors that cause this behavior include (1) shift of $[CO_3^{=}]/[HCO_3^{-}]$ with pH, (2) increased ionic strength, (3) formation of ion pairs between seawater cations and carbonate or bicarbonate. The carbonate ion pairs are stronger and tend to shift the balance of free ions toward bicarbonate. (From Plummer, L. N. 1975. *GSA Memoir* 142, Boulder Colorado; Geological Society of America p. 219–236.)

by saturating pure water with calcite at constant partial pressure of CO_2 from 10^{-3} to 1 atm. Small amounts of seawater result in undersaturation, but larger amounts in supersaturation.

BRINES

Thus far I have described only dilute solutions; the most concentrated was sea-water. Evaporation of natural waters can produce much more concentrated solutions, but the general quantitative theory that applies to dilute solutions is of little use. Not only are there ionic interactions of the type described by the Debye–Huckel theory and ionic interactions that can be described by ion-pairing equilibria, but in solutions more concentrated than 1 M specific ionic interactions occur that depend on all the major components of the solution. The result is that accurate prediction of ionic activities in brines requires a much larger field of empirical data than does the corresponding calculation in dilute solution.

For this reason, I will restrict myself to a qualitative discussion of the types of brines that may involve carbonate equilibria, and refer you to the literature for more details.*

When will carbonate equilibria be important in the formation of brines and evaporites? Figure 5.19 shows some of the possibilities depending on the amount of HCO_3^- relative to Ca^{++} and Mg^{++}. As long as $[HCO_3^-]$ is small compared to the concentration of divalent ions, the first mineral that precipitates is calcite (or a magnesian calcite; see p. 141), which removes the carbonate from solution. The remaining reactions involve precipitation of sulfates (e.g., gypsum, $CaSO_4 \cdot 2H_2O$, and epsomite, $MgSO_4 \cdot 7H_2O$) and halides (e.g., halite, NaCl, and carnallite, $MgCl_2 \cdot KCl \cdot 6H_2O$).

On the other hand, if HCO_3^- is in excess over Ca^{++} and Mg^{++}, precipitation of calcite and dolomite will remove the divalent ions from solution but will leave an excess of carbonate in solution. This then opens the possibility of precipitating more soluble carbonate minerals: nahcolite ($NaHCO_3$), natron ($Na_2CO_3 \cdot 10H_2O$), trona ($NaHCO_3 \cdot Na_2CO_3 \cdot 2H_2O$), thermonatrite ($Na_2CO_3 \cdot H_2O$), gaylussite ($Na_2CO_3 \cdot CaCO_3 \cdot 5H_2O$), pirssonite ($Na_2CO_3 \cdot CaCO_3 \cdot 2H_2O$), shortite ($Na_2CO_3 \cdot 2CaCO_3$), northupite ($Na_2CO_3 \cdot MgCO_3 \cdot NaCl$), hanksite ($9Na_2SO_4 \cdot 2NaCO_3 \cdot KCl$), tychite ($Na_2CO_3 \cdot MgCO_3 \cdot Na_2SO_4$), dawsonite ($NaAlCO_3(OH)_2$), etc.

Which minerals will precipitate? Qualitatively, it is clear that a solution with a high Na/Ca ratio will tend to produce carbonate minerals with only Na or with a high ratio of Na/Ca; solutions with relatively high Mg/Ca ratios will tend to precipitate northupite or tychite rather than gaylussite or pirssonite; but the precise concentration boundary where these salts precipitate cannot be predicted without a lot of laboratory data on solutions close in composition to those found in nature. Even so, equilibrium is not always reached. For example, equilibrium

---

* Eugster, H. P. 1980. *Ann. Rev. Earth Planetary Sci.* 8:35–63; Holland, H. D., op. cit., pp. 201–211.

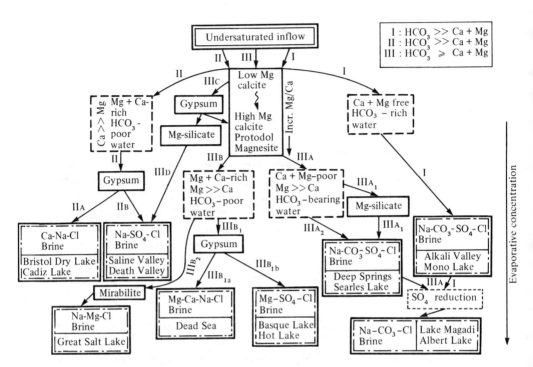

Fig. 5.19. Possible pathways for brine evolution, depending on the ratio of bicarbonate to Ca + Mg in the initial water. (From Eugster, H. P. 1980. Reproduced, with permission, from *Ann. Rev. Earth Plan. Sci.* 8:35–63. © 1980 by Annual Reviews Inc.)

between solution and solid phases that are covered with new solid phases may be attained slowly, if at all. Even the classical problem of which solids crystallize when seawater is evaporated, which was answered in part by van't Hoff in the early days of thermodynamic research (1896–1908), is still an open question today among geologists who would like to explain the particular sequence of minerals observed in marine evaporite deposits (Table 5.9).*

One simple example is shown in Fig. 5.20, which displays the ratio Na/Cl in waters associated with Lake Magadi in Kenya, East Africa. At low chloride concentrations, in the dilute surface water inflow, feldspar weathering introduces Na but not Cl, so the ratio Na/Cl increases as the water becomes more saline. When $[Cl^-]$ exceeds about 10^{-3} M, however, the ratio becomes more or less constant, and is similar for ground and surface waters, even up to the concentrations (0.1 to 1 M) found in saline springs and brines. The most concentrated brines, however, show a decrease in Na/Cl because of the precipitation of the mineral trona ($NaHCO_3 \cdot Na_2CO_3 \cdot 2H_2O$), which removes Na but not Cl.

---

* Harvie, C. E., Weare, J. H., Hardie, L. A., and Eugster, H. P. 1980. *Science* 208:498–500.

Table 5.9
Sequence of minerals formed by equilibrium evaporation of seawater at 25°C

Calculated*	Observed[†]
Calcite CaCO$_3$	
Gypsum CaSO$_4 \cdot$2H$_2$O	
Anhydrite CaSO$_4$	anhydrite
Anhydrite (A) + halite (H) NaCl	A + H
Glauberite CaSO$_4 \cdot$Na$_2$SO$_4$ + A + H	glauberite + A + H
Polyhalite (Po) 2CaSO$_4 \cdot$MgSO$_4 \cdot$K$_2$SO$_4 \cdot$2H$_2$O + A + H	Po + A + H
Epsomite MgSO$_4 \cdot$7H$_2$O + Po + A + H	
Hexahydrite MgSO$_4 \cdot$6H$_2$O + Po + A + H	
Kieserite (Ki) MgSO$_4 \cdot$H$_2$O + Po + A + H	Ki + Po + A + H
Carnallite (Car) KCl$\cdot$MgCl$_2 \cdot$6H$_2$O + Ki + Po + A + H	Car + Ki + Po + A + H
Car + Ki + A + H	Car + Ki + A + H
Bischofite MgCl$_2 \cdot$6H$_2$O + Car + Ki + A + H	

* Harvie et al., 1980. *Science* 208:498–500; also Harvie et al., 1979. *Geol. Soc. Am. Abstr. Programs* 11:440.

[†] Riedel, O. 1913. *Z. Kristallogr.* 50:139 (characterization of the Zechstein II sequence of evaporites).

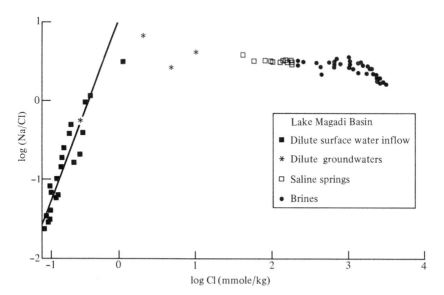

Fig. 5.20. Na/Cl ratio in waters from the Lake Magadi basin. The increase at low Cl is attributed to feldspar weathering, which introduces Na$^+$ but not Cl$^-$. In the intermediate range, brine concentration keeps Na/Cl constant; and in the most concentrated brines, precipitation of trona (NaHCO$_3 \cdot$Na$_2$CO$_3 \cdot$2H$_2$O) removes Na$^+$ but not Cl$^-$. (From Eugster, op. cit., p. 49.)

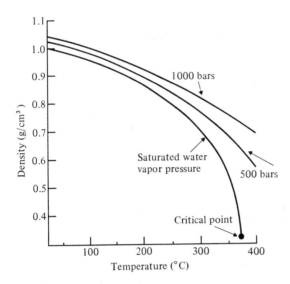

Fig. 5.21.   Density of water at saturated vapor pressures and at 500 and 1000 bars. The saturated water vapor pressure curve is given in numerical detail in the standard Steam Tables. (From Ellis, A. J. and Mahon, W. A. J. 1977. *Chemistry and Geothermal Systems.* New York: Academic Press, p. 119.)

HYDROTHERMAL SOLUTIONS

Temperature and pressure, as well as concentration of solutes, affect the activity of ionic species in solution. Although thermodynamic measurements at thousands of atmospheres and many hundreds of degrees Celsius are difficult, they have important applications in understanding the mechanisms of geothermal activity and those by which ore bodies are formed. Above, I have briefly touched on high pressure in connection with seawater but, as usual, I can only give you a few glimpses of a field that is dominated by empirical data and where the major questions involve the reconstruction of past sequences of temperatures and pressures.*

At constant pressure, the density of water decreases with increasing temperature, and increases with increasing pressure, as shown in Fig. 5.21. The lowest curve on Fig. 5.21 does not correspond to constant pressure, but to the saturation pressure at which liquid and gaseous water are in equilibrium. Along this line, the

---

* Two useful reviews are: Ellis, A. J. and Mahon, W. A. J. 1977. *Chemistry and Geothermal Systems.* New York: Academic Press; Barnes, H. L., ed. 1979. *Geochemistry of Hydrothermal Ore Deposits,* 2nd ed. New York: Wiley–Interscience.

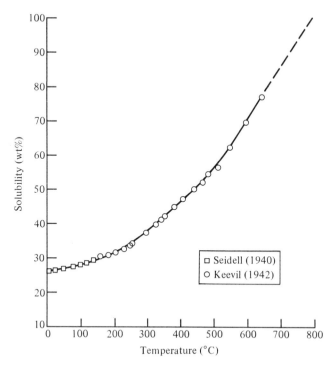

Fig. 5.22. Solubility of NaCl in water in equilibrium with water vapor from 0 to 800°C. The melting point of NaCl is 801°C, and at that point the liquid phase is essentially pure NaCl. (From Barnes, H. L. 1979. *Geochemistry of Hydrothermal Ore Deposits.* New York: Wiley-Interscience, p. 462.)

density of liquid water decreases with increasing temperature, and the density of water vapor increases. At the critical point (374.15°C, 221.29 bars*), the densities of liquid and gas are equal (0.322 g/cm³), and above this temperature there is only a single phase.

The shape of the saturation curve and the position of the critical point depend strongly on the concentration of solutes. For example, aqueous solutions saturated with NaCl retain the distinction between liquid and vapor to above 650°C, and the solubility of NaCl increases with temperature (Fig. 5.22). The vapor pressure of the NaCl-saturated solutions (Fig. 5.23) goes through a maximum (390 bars) at about 600°C, decreasing at higher temperatures as the solution becomes more like NaCl than like H_2O.

---

* 1 bar is defined as 10^6 dyne/cm² and is equal to 0.9869 atm or 750 Torr.

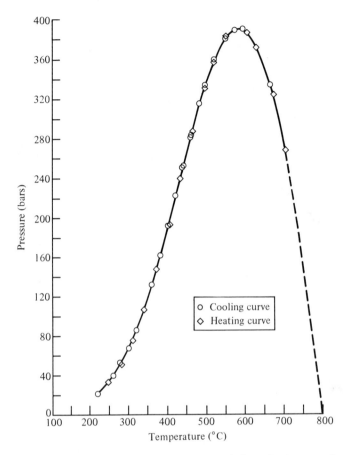

Fig. 5.23.   Vapor pressure of liquid consisting of mixtures of NaCl and H_2O at various temperatures. The curve is extrapolated to zero vapor pressure at 801°C, the melting point of NaCl..(From Barnes, op. cit., p. 463., after Sourirajan, S. and Kennedy, G. C. 1962. *Am. J. Sci.* 260:115–141.)

The solubility of a gas such as CO_2, on the other hand, decreases as the temperature increases to about 150°C, then increases with increasing temperature and becomes infinite at the critical point of the mixture. Addition of NaCl to the CO_2–H_2O system decreases the solubility of CO_2 but does not change the qualitative behavior. These effects are shown in Fig. 5.24, a plot of the familiar Henry's Law constant $K_H = [CO_2]/P_{CO_2}$. Replacing pressure by fugacity, which corrects for the nonideality of the gas phase, reduces the data scatter somewhat.

Acid–base and solubility equilibria also change with temperature and pressure, as you saw in the discussion of deep-sea equilibria. A more extended range of

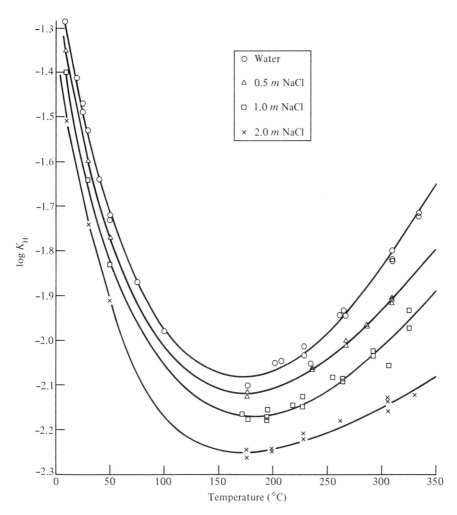

Fig. 5.24. Henry's Law constant K_H (mole/Kg atm) for CO_2 solubility in water and NaCl solutions as a function of temperature. These data are summarized in Table 2.4. Part of the upper curve corresponds to data listed in Table 2.2 for 0°C to 200°C. (Data from Ellis, A. J. and Golding, R. M. 1963. *Am. J. Sci.* 261:47–60.)

temperature is covered in Fig. 5.25; the pressure varies with temperature so as to be equal to the saturated vapor pressure of water (see Fig. 5.21), which is 1 bar at 100°C and 220 bars at 374°C. All the graphs of Fig. 5.25 show upward curvature; except for H_2O and NH_4^+, pK increases at sufficiently high temperature. It is clear, however, that at least two and probably more parameters are required to describe the changes illustrated.

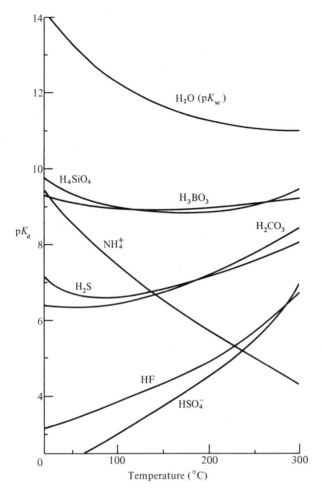

Fig. 5.25.   Variation of acidity constants with temperature at saturated water vapor pressures (1 bar at 100°C, 10 bars at 180°C, 30 bars at 234°C, and 86 bars at 300°C). (From Ellis and Mahon, op. cit., p. 128.)

These parameters are often expressed as thermodynamic functions. For example, the pressure dependence of an equilibrium constant K is given by the change in partial molar volume between reactants and products:

$$\Delta V^0 = \sum_{\text{prod}} V_i - \sum_{\text{react}} V_j \quad \text{(at standard conditions)},$$

$$\left(\frac{\partial \ln K}{\partial P} \right)_T = -\frac{\Delta V^0}{RT},$$

where R is the gas constant (1.987 cal/mol·deg or 82.06 cm^3·atm/mol·deg) and T is the temperature in °K. In general, ΔV^0 is a weak function of temperature, pressure, and solution composition.* The temperature dependence is given in terms of the heat capacity (C_P) and enthalpy (H) change:

$$\Delta C_P^0 = \sum_{\text{prod}} C_{Pi} - \sum_{\text{react}} C_{Pj},$$

$$\left(\frac{\partial \Delta H^0}{\partial T}\right)_P = \Delta C_P^0,$$

$$\left(\frac{\partial \ln K}{\partial T}\right)_P = \frac{\Delta H^0}{RT^2}.$$

In general, ΔC_P^0 is a weak function of temperature, pressure, and solution composition, as is ΔH^0. These relationships (with ΔC_P^0 and ΔV^0 constant) lead to the integrated form

$$\ln K = B - \frac{\Delta H^0}{RT} - \frac{\Delta C_P^0}{R} \ln T - \frac{\Delta V^0}{RT} (P - 1), \qquad (5.63)$$

where B is a constant that can be evaluated if K is known at any condition for which T and P are known. More complicated forms, with more parameters, can be obtained by expressing ΔV^0 and ΔC_P^0 in power series.

These thermodynamic equations are most useful for estimation or interpolation of empirical data, especially if the functions ΔC_P^0, ΔH^0, ΔV^0 can be found in standard tables of thermodynamic quantities. However, such tables rarely give data for complex mixtures, and as a result the prediction of equilibrium constants and related quantities often contains a good bit of guesswork.

It is unusual to find measurements of pH made at the high temperatures and pressures where solubilities are determined.[†] Sometimes a sample of a hydrothermal solution is taken and its pH is measured at room temperature and 1 atm pressure, but this measurement is only qualitatively related to the activity of hydrogen ion at the high temperatures and pressures of interest. Hence pH is mainly used as an unmeasured intermediate variable in chemical models of solubility processes. Figure 5.26 gives data for the ionization constants of CO_2 as a function of temperatures and pressure.

Since the focus of this book is on CO_2, $CaCO_3$, and related equilibria, the next subject to examine is how the solubility of calcite varies in the hydrothermal region. Figure 5.27 is a plot of calcite solubility at 150°C as a function of CO_2

---

* For more details see Lewis, G. N. and Randall, M. 1961. *Thermodynamics*, 2nd ed., revised by K. Pitzer and L. Brewer. New York: McGraw-Hill; Stumm and Morgan, op. cit., Chapter 2; Pytkowicz, R. M., ed. 1979. *Activity Coefficients in Electrolyte Solutions*. Boca Raton, Fla: CRC Press; or other thermodynamics books.

[†] Disteche, A. 1974. In *The Sea*, vol. 5, E. D. Goldberg, ed. New York: Wiley, Chapter 2.

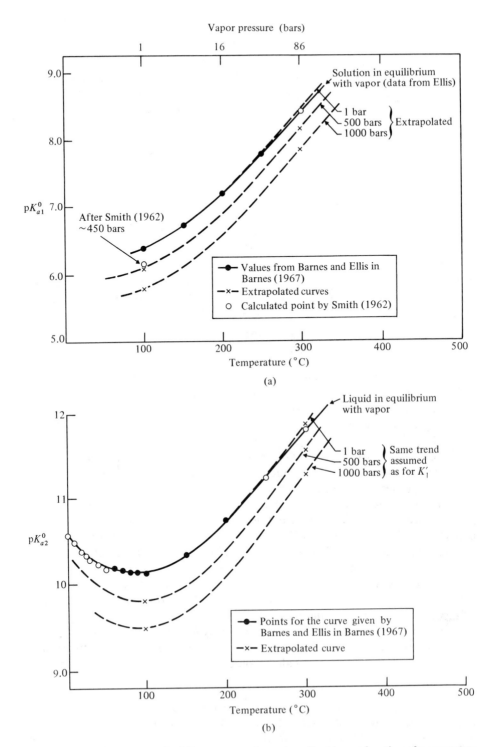

Fig. 5.26. First and second acidity constants for carbon dioxide as a function of temperature and pressure. Data corresponding to the temperature range below 200°C can be found in Table 2.2. (From H. D. Holland, private communication.)

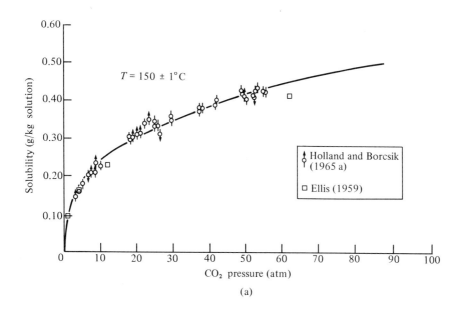

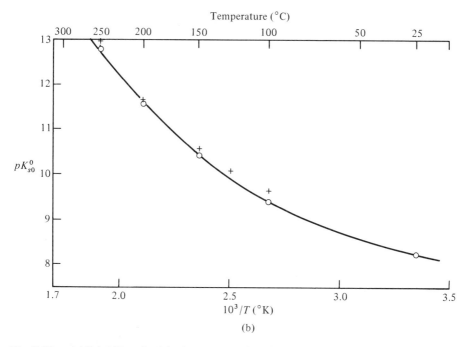

Fig. 5.27. (a) Solubility of calcite in water as a function of CO_2 pressure in the vapor phase at 150°C. The curve is calculated from an equation of the form of (5.17). (From Barnes, op. cit., p. 479, after Holland, H. D., and Borcsik, M., 1965. Symposium, *Problems of Postmagmatic Ore Deposition*, Prague, 2:418–421.) (b) Solubility product of calcite as a function of temperature. (⊙) Ellis, A. J., 1959. *Am. J. Sci.* 257:354–365, (+) Ellis, A. J., 1963. *Am. J. Sci.* 261: 259–267.

pressure up to 60 atm. The shape of the curve can be accurately predicted by the same simple chemical model used for river water at the beginning of this chapter. If the only components of the system are CO_2, $CaCO_3$, and H_2O, then the alkalinity (which is equal to $2[Ca^{++}]$) is given by (see pp. 77–78 and 113)

$$\log A = \frac{1}{3}(pK_{a2}^0 - pK_{a1}^0 - pK_H^0 - pK_{s0}^0 + \log P + \log 2) + f(I) + \cdots \qquad (5.17)$$

or

$$[Ca^{++}] = \left(\frac{K_{a1}^0 K_H^0 K_{s0}^0 P}{4K_{a2}^0 \gamma_-^2 \gamma_{++}}\right)^{1/3},$$

and hence the solubility (which is equal to $[Ca^{++}]$) should be proportional to $P^{1/3}$. This is demonstrated by the good fit of the curve (representing Eq. 5.17) to the data in Fig. 5.27a. Each isotherm like this gives a value for K_{s0} (given the other constants—see Table 2.2, Fig. 5.24, and Fig. 5.26) for a particular temperature. A few of these values are displayed in Fig. 5.27b and listed in Table 2.2.

The effect of temperature on the solubility of calcite at constant pressure can be calculated from Eq. 5.17, and these curves are compared with experimental data in Fig. 5.28. Note that within the pressure range and temperature range considered, the variation of activity coefficients (Fig. 5.29) is only a few percent, and that the major factors determining solubility are the direct effect of CO_2 partial pressure

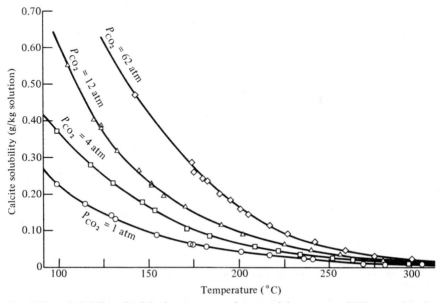

Fig. 5.28.   Solubility of calcite in water at various partial pressures of CO_2 as a function of temperature. (From Barnes, op. cit., p. 480, after Ellis, op. cit.)

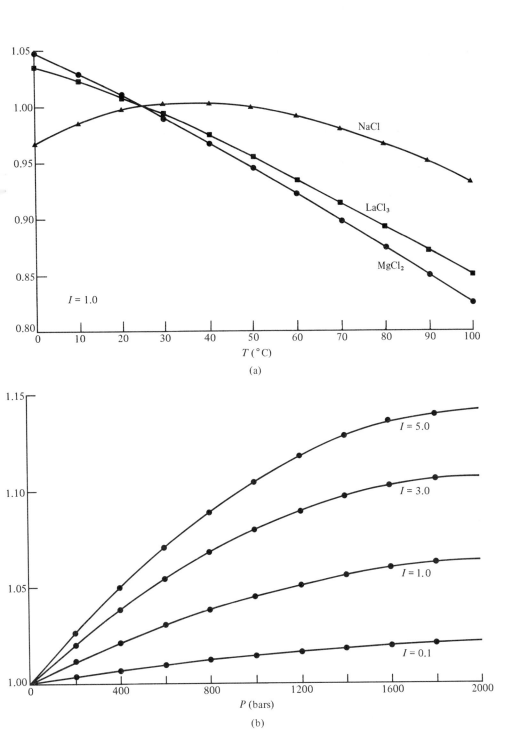

Fig. 5.29. (a) Temperature dependence of mean activity coefficient for three salts of different charge type. (b) Pressure dependence of the mean activity coefficient of NaCl at 25°C and ionic strengths up to 5.0 mole/L. (From Pytkowicz, R. M., ed. 1979. *Activity Coefficients in Electrolyte Solutions.* Boca Raton, Fla: CRC Press, pp. 81 and 112.)

(Eq. 5.17) together with the temperature dependence of the four equilibrium constants (Figs. 5.24, 5.26, and 5.27). Although the solubility product increases with increasing temperature, this is more than offset by the decrease in CO_2 solubility. The net result is that hydrothermal solutions, saturated with CO_2 and calcite at high temperatures, become capable of dissolving more limestone rock as they rise toward the surface and are simultaneously cooled and reduced in pressure. Ultimately, these solutions lose CO_2 by evaporation and deposit calcite.

Addition of NaCl to solutions saturated with CO_2 and $CaCO_3$ increases the solubility of $CaCO_3$ (Fig. 5.30a). The effect, expressed as the ratio of solubility in

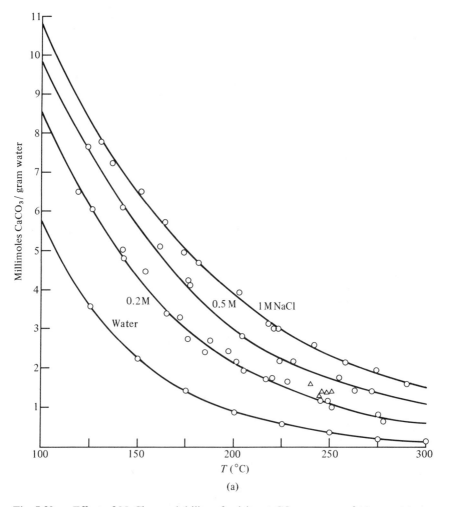

Fig. 5.30.   Effect of NaCl on solubility of calcite at CO_2 pressure of 12 atm: (a) temperature dependence at several constant NaCl concentrations (from Ellis, A. J. 1963. *Am. J. Sci.* 261:259–267).

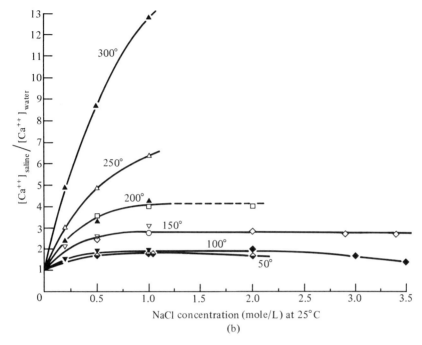

Fig. 5.30. (b) ratio of calcite solubility in the presence of NaCl to that in water as a function of NaCl concentration at several constant temperatures (from Barnes, op. cit.; data from Holland and Borcsik, op. cit. and Ellis, op. cit.).

saline water to that in pure water (Fig. 5.30b), is greater at high temperatures. At temperatures below 100°C, this ratio goes through a broad maximum between 0.5 and 3.0 M, which can be attributed to the effect of ionic strength on the activity coefficients in Eq. (5.17): both γ_- and γ_{++} for unassociated ions decrease as ionic strength increases to about 1 to 3 M, then increase (recall the Davies equation (2.36), and Fig. 2.5). Neither pressure nor temperature has a strong effect on $\gamma_\pm$ (Fig. 5.29). In addition, there are ion pairs $NaHCO_3$ and $NaCO_3^-$, (well documented in Tables 5.6 and 5.7), whose formation will tend to increase the solubility of $CaCO_3$ as NaCl concentration is increased.

One way of expressing the ion-pairing effect within the form of Eq. (5.17) is to note that the effective value of γ_- will be decreased by any ion (such as Na^+) associated with HCO_3^-. On the other hand, since $CaCl^+$ is a weak complex, the addition of NaCl would not be expected to affect γ_{++}.

The increase in solubility due to moderate concentrations of NaCl (as shown in Fig. 5.30) is not sufficient to offset the steep decrease in solubility due to increasing temperature alone, but it does indicate that saline waters will tend to be less aggressive in dissolving limestone rock than waters containing only CO_2 and $CaCO_3$.

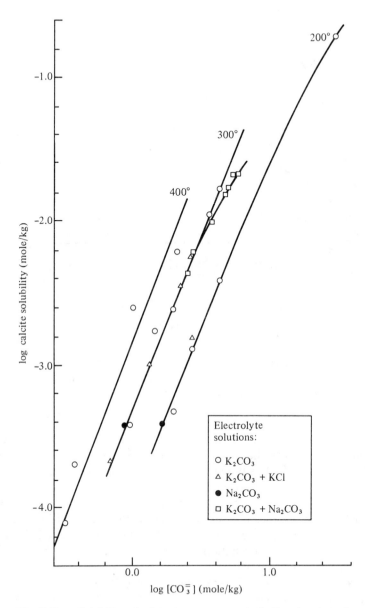

Fig. 5.31.   Solubility of calcite in concentrated alkali carbonate solutions. The steep positive slope implies formation of complexes such as $Ca(CO_3)_4^{6-}$. (From Barnes, op. cit.; data from Malinin, S. D. and Dernov–Pegarev, V. F., 1974. Geokhimiya 3:454–462)

Ion-pairing effects may be dominant in extremely high salinities and in alkaline solutions. For example, Fig. 5.31 shows how the solubility of $CaCO_3$ varies with temperature and with carbonate concentration in solutions of alkali carbonates. Note first that, as temperature decreases from 300 to 200°C, at carbonate concentrations between 1 and 3 mole/kg, the solubility of $CaCO_3$ decreases by about a factor of 5. In contrast (see Fig. 5.28), the solubility of $CaCO_3$ in pure water at 62 atm *increases* by about a factor of 10 for the same temperature decrease. Part of this dramatic difference is a result of CO_2 solubility (which increases strongly with decreasing temperature in the data of Fig. 5.28), in contrast with the fixed carbonate concentration of the alkaline media shown in Fig. 5.31.

Another important difference is the ion pairing. If the predominant ions in the alkaline solutions were Ca^{++} and $CO_3^=$, the solubility would be expected to decrease with increasing carbonate content, according to the solubility product:

$$S = [Ca^{++}] + \cdots = \frac{K_{s0}}{[CO_3^=]} + \cdots .$$

This is clearly not the case in Fig. 5.31. The lines have slope $+3$ instead of -1, and imply that the dominant dissolved calcium species is not Ca^{++} but a complex with four carbonates:

$$S = [Ca(CO_3)_4^{6-}] + \cdots = K_{s4}[CO_3^=]^3 + \cdots .$$

Presumably, other complexes such as $Ca(CO_3)_3^{4-}$ and $Ca(CO_3)_2^=$ are also present, and if data were available to sufficiently low carbonate concentrations, the full curve would have the U shape characteristic of the solubility of salts in solutions of excess complexing anion.*

PROBLEMS

1. Rainwater collected near the seacoast in an isolated part of the world was analyzed and found to have the composition

   $[Na^+] = 48 \cdot 10^{-3} \, M$, $[SO_4^=] = 3 \cdot 10^{-3}$, $[K^+] = 1 \cdot 10^{-3}$,

   $[Cl^-] = 55 \cdot 10^{-3}$, $[Mg^{++}] = 5 \cdot 10^{-3}$, $[Ca^{++}] = 1 \cdot 10^{-3}$.

   These ionic concentrations are in approximately the same ratio as seawater (Table 5.4), except 10 times as dilute. If the atmospheric CO_2 is $10^{-3.5}$ atm, calculate the pH and alkalinity of the rainwater.

   Answer: Note that the alkalinity of seawater is approximately $2.3 \cdot 10^{-3}$, and so the seawater contribution to the rainwater alkalinity would be $0.23 \cdot 10^{-3}$. pH = 7.59.

---

* See Problem 63, and *Solubility and pH calculations*, pp. 26–27; see also *Ionic Equilibrium*, p. 278, and Stumm and Morgan, op. cit. pp. 248, 543.

2. Early workers on rain analysis measured total nitrogen but not $[NH_4^+]$ and $[NO_3^-]$ independently. Suppose you knew that

$$[\text{Total N}] = 42 \cdot 10^{-6}\,M, \qquad [SO_4^=] = 28 \cdot 10^{-6},$$
$$[Cl^-] = 6 \cdot 10^{-6}, \qquad [Na^+] = 3 \cdot 10^{-6},$$

and $pH = 4.20$. Make use of the charge balance to estimate $[NO_3^-]$ and $[NH_4^+]$ separately.

Answer: $[NH_4^+] = 19 \times 10^{-6}\,M$, $[NO_3^-] = 23 \times 10^{-6}\,M$

3. In an attempt to reconstruct historical records of precipitation acidity from analyses of rainwater, one method assumes that there is negligible discrepancy in the charge balance

$$\sum_i z_i[M_i] - \sum z_j[X_j],$$

where M_i is a cation with charge z_i and X_j is an anion with charge z_j (Liljestrand, H. and Morgan, J. J. 1979. *Tellus* 31:421–431). Suppose data were given for $[Na^+]$, $[K^+]$, $[Ca^{++}]$, $[Mg^{++}]$, $[NH_4^+]$, $[Cl^-]$, $[NO_3^-]$, $[SO_4^=]$, and alkalinity (or mineral acidity) A. Use the charge balance to derive an expression for pH. Use the data in Table 5.1 to estimate the ionic strength. If the rainwater contained an organic acid with $pK^a = 4.3$ at a concentration of $5 \cdot 10^{-6}\,M$, but this was neglected in the analysis, what error in pH would occur?

4. Granat (1972. *Tellus* 24:550–560) developed a *source strength* model for rainwater in which he assumed Na^+ and Cl^- to be contributed primarily by sea salt; K^+, Ca^{++}, and Mg^{++} to be contributed by soil dust; NH_4^+ to be derived from NH_3 sources; NO_3^- to be derived from HNO_3 sources; and $SO_4^=$ to be derived from H_2SO_4 (oxidized SO_2 pollution) sources. Therefore the strengths of these five sources and the partial pressure of atmospheric CO_2 determine the ionic concentrations, the charge balance, and the pH. Derive that relationship.

5. Recall that sea salt contains some $SO_4^=$ and Mg^{++} (Table 5.4) and assume that the soil is derived primarily from limestone and dolomite, so that each mole of Ca^{++} or Mg^{++} from the soil source contributes one mole of carbonate. Derive the modified relationship for pH (as in Problem 4) and estimate the error if the soil composition is unknown.

6. Frohliger and Kane (1975. *Science* 189:455–456) measured the acidity of rainwater by adding NaOH to the sample until $pH = 9$ was reached, and found that this was not consistent with the acidity calculated from the initial pH. They hypothesized the presence of weak acids ("bound H^+," nature unspecified). Derive equations for their titration acidity value A'' and the pH-derived acidity by using the rain composition in Table 5.1. What is the source of the discrepancy? Test your hypothesis against their published data: $pH = 5.62$, $A'' = 7.27 \cdot 10^{-5}$; $pH = 4.12$, $A'' = 19.0 \cdot 10^{-5}$; $pH = 4.82$, $A'' = 44.0 \cdot 10^{-5}$; etc.

Answer: CO_2, NH_4^+

7. The presence of weak acids in a mixture with strong acid can be detected by a Gran titration (Lee, Y. H. and Brosset, C. 1978. *Water, Air and Soil Pollution* 10:457–469). Suppose a rain sample contained $30 \cdot 10^{-6}$ mole/L of H_2SO_4 and $5 \cdot 10^{-6}$ mole/L of acetic acid ($pK_a = 4.75$). Calculate 10^{-pH} as a function of added base and compare the curves obtained with and without the acetic acid. Estimate the smallest concentration

of weak acid that could be detected in the presence of $30 \cdot 10^{-6}$ M H_2SO_4 if pH were measured with an accuracy of ± 0.01 unit.

8. Calculate the Gran titration function appropriate for a sample (titrated with base) containing $3 \cdot 10^{-5}$ M H_2SO_4, $1 \cdot 10^{-5}$ M CO_2, and $2 \cdot 10^{-5}$ M NH_4^+. Can these three components be evaluated accurately on the basis of a single titration?

9. Rainwater containing only CO_2 in equilibrium with the atmosphere ($P = 10^{-3.5}$) reacts with limestone ($CaCO_3$); at the point sampled, pH = 7.7. Estimate the alkalinity. Is the solution saturated with $CaCO_3$? How much more reaction is possible?

   Answer: $A = 10^{-3.62}$; since $[Ca^{++}][CO_3^=] = 10^{-10.17}$, the solution is undersaturated. At saturation (see Eq. 5.17), $A = 10^{-3.06}$, a factor of 3.6 times as large (provided P is constant).

10. If all the CO_2 in rainwater reacts with $CaCO_3$ to reach saturation, pH is given by Eq. (5.16). Derive the expression for akalinity if the final solution is still in equilibrium with the atmosphere (Eq. 5.17).

11. If rainwater contains mineral acidity (Table 5.1), it will be more aggressive in attacking limestone rock than rain consisting of only CO_2 and H_2O. Calculate the final pH and alkalinity if the water whose analysis is given in Table 5.1 is allowed to become saturated with $CaCO_3$ while remaining in equilibrium with the atmosphere.

    Answer: pH = 8.25, $A = 10^{-3.06}$.

12. Rainwater contacts limestone but flows into cracks where it is no longer in equilibrium with the atmosphere. How does this change the answers to Problems 10 and 11?

13. What effect do the ion pairs $CaHCO_3^+$, $CaCO_3$, and $CaOH^+$ have on the calculations of Problems 10 and 11?

14. Well No. 332 in Central Pennsylvania was drilled in a dolomite rock area, and a sample of water at 13°C had the following composition (in mg/kg or ppm): $[Ca^{++}]$, 56; $[Mg^{++}]$, 29; $[Na^+]$, 1.0; $[K^+]$, 0.8; $[HCO_3^-]$, 263; $[SO_4^=]$, 17; $[Cl^-]$, 4.7; $[NO_3^-]$, 27 (Langmuir, D. 1971. *Geochim. Cosmochim. Acta* 25:1023–1045). The water was saturated with oxygen (102% sat) and had a CO_2 content in equilibrium with an atmosphere containing 616 ppm CO_2. Determine whether the water is saturated with calcite, dolomite, or both.

    Answer: Both

15. In limestone caves, $CaCO_3$ precipitates or crystallizes in stalactites, stalagmites, etc., as saturated groundwater loses CO_2 to the cave atmosphere or evaporates within the cave. If groundwater initially in equilibrium with 0.2 atm CO_2 becomes saturated with calcite as it goes through the ground (constant D), calculate the amount of $CaCO_3$ deposited per liter of groundwater: (a) if the water comes into equilibrium with water-saturated cave air ($P_{CO_2} = 10^{-3.0}$ atm) at 10°C; (b) if the groundwater evaporates completely within the cave.

    Answer: This calculation is much more difficult than it looks. The ionic strength of the groundwater is about 10^{-2} M, so that activity coefficients are not negligible. Finding the pH (approximately 6.55) requires solution of a cubic polynomial. This is a good problem to try on the computer. Approximate answers are (a) 220 mg/L and (b) 300 mg/L. Compare with problems 5, 8, 22–26 in Chapter 4 and problem 18 below.

*16. A river draining a granitic terrain at a temperature of 25°C contains 3.6 mg Ca^{++} per kg H_2O (ppm), 0.8 ppm Mg^{++}, 1.0 ppm Na^+, and 0.5 ppm K^+. These cations are balanced by HCO_3^-. If this river is in equilibrium with atmospheric CO_2 ($3.2 \cdot 10^{-4}$ atm), what is its pH?

Answer: pH = 7.8

*17. The river of Problem 16 encounters a limestone consisting of nearly pure $CaCO_3$. Calculate the concentrations of Ca^{++} and HCO_3^-, as well as the pH of the river, after equilibration at the same temperature and pressure.

Answer: $[Ca^{++}] = 4.4 \cdot 10^{-4}$ M; $[HCO_3^-] = 1.0 \cdot 10^{-3}$ M; pH = 8.32

*18. The bed of the river of Problem 16 contains extensive fractures in the limestone. Water percolates down these fractures. If the water remains at equilibrium with limestone at 25°C but is no longer in equilibrium with the atmosphere while in these fractures, calculate the new concentrations of Ca^{++} and HCO_3^-, and the pH.

19. Test to see whether albite can be weathered to kaolinite (Eq. (5.20)) by average river water from Africa (Table 5.2).

20. Verify that the ionic strength of seawater is $700 \cdot 10^{-3}$ mole/kg and calculate activity coefficients for Ca^{++} and $CO_3^=$ by using the Davies equation. Test to see if K_{s0}^0 is exceeded at 25°C. Compare with K_{s0} as listed in Table 5.5.

21. Buch and coworkers (1932. *Rapp Cons. Explor. Mer.* 79:1) measured alkalinity, pH, and P_{CO_2}, and determined the first ionization constant of carbon dioxide in seawater by means of the equation

$$K'_{1(b)} = \frac{A(10^{-pH})}{\alpha_0 P_{CO_2}(1 + 2K'_2 \cdot 10^{+pH})a_{H_2O}},$$

where A is carbonate alkalinity (total alkalinity corrected for borate etc.), α_0 is the solubility of CO_2 (mole/L) in pure water at CO_2 pressure of 1 atm, and a_{H_2O} is the activity of water on a scale where a_{H_2O} approaches 1.0 in pure water. Find the relationship between $K'_{1(b)}$ and K'_1 as defined in the text (Skirrow, op. cit., p. 43).

22. Buch and coworkers (op. cit.) defined the first ionization constant of carbon dioxide in seawater to be

$$K'_{1(b)} = \frac{10^{-pH}[HCO_3^-]_T}{a_{H_2O} \cdot a_{CO_2}},$$

where a_{H_2O} is the activity of water on a scale where a_{H_2O} approaches 1.0 in pure water, and a_{CO_2} is the activity of CO_2 determined by the partial pressure through Henry's Law in the form

$$a_{CO_2} = \alpha_0 P_{CO_2},$$

where α_0 is the solubility of CO_2 in pure water at 1 atm partial pressure. Show the relationship of $K_{1(b)}$ to Lyman's $K'_{1(L)}$ (Problem 23) and to the constants K_{a1} and K_H as defined in the text. What happens when the salinity approaches zero? (See Skirrow, op. cit., p. 40.)

23. Lyman (1956, Ph.D. thesis, University of California at Los Angeles) defined the first ionization constant of carbon dioxide in seawater to be

$$K'_{1(L)} = \frac{10^{-pH}[HCO_3^-]_T}{[CO_2]_{aq} + [H_2CO_3]},$$

---

* These problems were courteously provided by Prof. H. D. Holland.

where pH is measured on the NBS scale and $[HCO_3^-]_T$ is the sum of free $[HCO_3^-]$ and all its ion pairs with cations such as Na^+, Mg^{++}, Ca^{++}, etc. If the equilibrium constant for protonation of dissolved CO_2 at 25°C is given by

$$K_0 = \frac{[H_2CO_3]}{[CO_2]_{aq}} = 10^{-2.85},$$

would it be independent of ionic medium? Find the relationship between $K'_{1(L)}$ and the first acidity constant as defined in Chapter 2:

$$K_{a1} = \frac{[H^+][HCO_3^-]}{[CO_2]}.$$

What happens when the ionic medium approaches pure water (salinity approaches zero)? (See Skirrow, op. cit., p. 39.)

24. Calculate γ_+ for H^+ from the ratio of "hybrid NBS scale" constant to "Hansson" constants in Table 5.5. Compare with an estimate via the Davies equation. Why are these values not all identical?

25. A simplified ion-pair model for seawater might include the equilibria

$$[MgSO_4] = 10^{+1.04}[Mg^{++}][SO_4^-], \qquad [NaSO_4^-] = 10^{-0.07}[Na^+][SO_4^-],$$

with $[MgCl^+]$ and $[NaCl]$ negligible and the total concentrations given by Table 5.4. Compare the results of this model with Table 5.6 or 5.7.

26. The maximum buffer capacity $\beta = (\partial A/\partial pH)_{C_T}$ for seawater is developed close to pH = 6.0 and pH = 9.0. Show that β_{max} is approximately 0.58 times the total carbonate concentration. Find the buffer capacity at pH = 8.0 and 25°C.

    Answer: $4 \cdot 10^{-4}$ mole/L (Skirrow, op. cit., p. 11)

27. Show that the buffer capacity $\beta' = (\partial C_T/\partial pH)_A$ differs little from β as defined in the text under most ocean surface conditions. Find the relationship between the two quantities.

    Answer: $\beta'/\beta = -C_T/A$

28. Culberson, Pytkowicz, and Hawley (1970. *J. Mar. Res.* 28:15–21) proposed a method for alkalinity determination that uses $[H^+]_{(T)} = 10^{-pH}/\gamma_{H(T)}$ in a mass balance equation. The "total" activity coefficient is evaluated from the equation

$$\gamma_{H(T)} = \frac{[H^+]_f + [HSO_4^-] + [HF]}{10^{-pH}},$$

where $[H^+]_f$ is the concentration of free hydrogen ions calculated from a chemical model that includes the possibility of protonating sulfate and fluoride as well as water. Derive an equation for $[H^+]_T$ as a function of pH, total sulfate, and total fluoride, and evaluate it for a typical seawater (Table 5.4) at pH = 4.5.

29. The alkalinity values measured by Edmond, who used Gran's potentiometric titration method (as described in Chapter 3), agreed within 1% with values measured by means of a pH method originally proposed by Anderson and Robinson (1946, *Ind. Eng. Chem.* 18:767–769) and modified by Culberson et al. (op. cit., Problem 28) (data are given in Figs. 5.3 to 5.6). In this latter method, excess HCl is added to the seawater, the carbon dioxide is driven off by purging for 5 min with water-saturated air, and the final pH is measured. The total alkalinity is obtained by assuming a hydrogen ion activity coefficient

γ_+ appropriate for the pH scale used, and finding the excess acid from the definition $[H^+]_{xs} = 10^{-pH}/\gamma_+$. (Here $\gamma_+ = 0.741$ if the pH scale is calibrated by setting pH = 4.007 in a 0.05 molal potassium acid phthalate buffer. This is very close to the NBS scale, but high compared with $\gamma_+ = 0.69$ from the Davies equation.) If V mL of HCl with concentration C are added to a sample of volume V_0, show that

$$A = \frac{CV - [H^+]_{xs}(V + V_0)}{V_0}.$$

30. By using the model of Table 5.7, show that titration of seawater with HCl, so that all carbonate species (including ion pairs with cations) are converted to CO_2, increases Na^+ by 0.04%, Mg^{++} by 1.3%, and Ca^{++} by about 10%. Show the effect of this process on the shape of the alkalinity titration curve.

31. Phosphate, being a triprotic base, will take up protons as pH is lowered during an alkalinity titration. Calculate the phosphate contribution to the total alkalinity of seawater if the total phosphate concentration is $2 \cdot 10^{-6}$ M. What is the maximum effect on V/V_0 over the pH range 8.0 to 4.5? (See Eq. (3.54) and Fig. 3.7.)

32. Oceanographers tend to adhere to a convention where total alkalinity is corrected for borate and silicate but not sulfate. Calculate the contribution from sulfate in a brackish water with half the salinity of seawater but the same relative ionic composition if the alkalinity titration is carried to an end point at pH = 4.5.

33. If HSO_4^- were neglected, but not H_3PO_4 (this occurs in the literature in some definitions of alkalinity), what effect does this have on the difference between A_T and A_c for seawater (Table 5.4)? Use typical values for seawater concentrations.

34. In the Gran method for determining alkalinity (Fig. 3.7) titration data are taken at pH values as low as 3, for extrapolation back to the end point. Derive an equation for the region of the titration curve between pH = 3 and pH = 4.5 that includes the effect of sulfate, and compare with Eq. (3.48). Use the data of Table 3.1 to estimate the error in extrapolation to obtain V_2. Is this distinguishable from the error due to estimating γ_+, the activity coefficient of H^+?

35. Derive the equation for the homogeneous (Revelle) buffer factor (Eq. (5.40)), including the effect of boron on alkalinity. Compare with Eq. (5.49).

36. For seawater with total alkalinity $2.67 \cdot 10^{-3}$ mole/kg, total dissolved silicon concentration $1.5 \cdot 10^{-4}$ mole/kg, total dissolved phosphorus $3.0 \cdot 10^{-6}$ mole/kg, and pH = 8, calculate the contribution of $[SiO(OH)_3^-]$, $[HPO_4^=]$, and $PO_4^{\equiv}$ to the total alkalinity ($pK_a' = 9.4$ for Si, $pK_{a2}' = 6.06$ for P, $pK_{a3}' = 8.56$).

37. Edmond (1970. *Deep Sea Res.* 17:737–750) states that the major species accounting for total alkalinity in seawater make contributions in the proportion

$$HCO_3^- \quad 89.8\%, \qquad B(OH)_4^- \quad 2.9\%, \qquad HPO_4^= \quad 0.1\%,$$
$$CO_3^= \quad 6.7\%, \qquad SiO(OH)_3^- \quad 0.2\%, \qquad PO_4^{\equiv} \quad 0.05\%.$$

To what total concentrations of B, Si, and P do these values correspond if pH = 8.0 and $A_T = 2.21 \cdot 10^{-3}$ Eq/kg?

38. Suppose the acid rain, whose composition is given in Table 5.1, underwent anaerobic digestion, which converted $SO_4^=$ to HS^- and NO_3^- to NH_4^+. The reduction would probably be accomplished by microbes who would convert carbohydrate (roughly CH_2O)

to CO_2 in order to provide electrons. You can therefore make approximate balanced equations for the reduction of $SO_4^=$ and NO_3^- by CH_2O:

$$SO_4^= + 2CH_2O \rightarrow HS^- + HCO_3^- + CO_2 + 2H_2O,$$
$$NO_3^- + 2CH_2O \rightarrow NH_4^+ + 2HCO_3^- + H_2O.$$

Predict the alkalinity (or mineral acidity) and pH of the result. What effect do HS^- and NH_4^+ have on the total alkalinity?

39. Show that the contribution of $[MgOH^+]$ to the total alkalinity is approximately equal to that for $[OH^-]$. Show that the ratio of these two contributions remains constant as pH changes, even though their magnitudes are strong functions of pH.

40. What effect do the ion pairs $NaSO_4^-$, $CaSO_4$, and $MgSO_4$ have on the calculation of $[HSO_4^-]'$ as a correction for the alkalinity?

41. To water of pH $= 7.2$ and alkalinity $1.0 \cdot 10^{-3}$ is added an equal quantity of a monoprotic organic acid with $pK_a = 5.4$ in concentration $1.0 \cdot 10^{-4}$ mole/L. Calculate the pH of the resulting mixture.

42. If the mixture of Problem 41 were titrated with a strong acid, to the same endpoint as would be expected if there were only carbon dioxide, what alkalinity value would be obtained?

43. If the organic acid were actually added as the disodium salt of phthalic acid (with $pK_{a1} = 2.95$ and $pK_{a2} = 5.40$), how would this affect the answers to Problems 41 and 42?

44. If Problems 41–43 were carried out in seawater, how would the results be affected?

45. Use the data in Tables 5.5 and 5.7 to show that surface seawater is supersaturated with $CaCO_3$. What would be the concentrations of Ca^{++} and total carbonate if saturation equilibrium with $CaCO_3$ were reached at pH $= 8.0$?

46. Show that if seawater pH were raised to 9.0 and equilibrium were established with solid calcite, that C_T would be reduced by a factor of about 20, whereas $[Ca^{++}]$ would be reduced by only about 20%.

47. Using data from Figs. 5.3, 5.4, 5.8, and 5.9, plot the degree of saturation with $CaCO_3$ of seawater to a depth of 6000 m. Assume the temperature profile as follows (N. Pacific average):

$$0\,m \quad 20°C, \qquad 600\,m \quad 6°C, \qquad 2000\,m \quad 2°C,$$
$$200\,m \quad 10°C, \qquad 1000\,m \quad 3°C, \qquad >2000\,m \quad 2°C.$$

Estimate the lysocline depth and compare with Fig. 5.10.

48. Calcite was equilibrated with seawater at 25.0°C while a nitrogen–CO_2 gas mixture (18.6% CO_2) was bubbled through the reaction vessel at a total pressure of 1 atm. The salinity was measured to be 34.77‰, so that $[Ca^{++}]$ could be calculated from the composition of normal seawater (35.00‰; Table 5.4). The steady pH value obtained was 6.53. Calculate the concentration solubility product of calcite in seawater.

49. In shallow waters with $CaCO_3$ sediments, more CO_2 can be absorbed from the atmosphere than in the open ocean because of rapid reaction with suspended magnesian calcite particles (see p. 141):

$$Mg_xCa_yCO_3(s) + CO_2 + H_2O \rightleftharpoons xMg^{++} + yCa^{++} + 2HCO_3^-,$$

where x is approximately 0.08 and $x + y = 1$ (see Wollast et al. op. cit.). Since HCO_3^- is produced, alkalinity is not constant, but $D = 2(C_T - [Mg^{++}] - [Ca^{++}])$ would be unaffected by this reaction. The ion product $[Ca^{++}][CO_3^=]\gamma_{++}\gamma_= = 10^{-8.2}$ for this solid, compared to $10^{-8.5}$ for pure calcite. Estimate the value of the buffer factor $(\partial \log P)/(\partial \log C_T)_D$ and compare with Eq. 5.48.

50. F. McIntyre (1978. *Thalassia Jugoslavica* 14:63–98) has proposed a "minimal model" of the atmosphere–ocean response to increasing anthropogenic CO_2 (Fig. 5.11). The terrestrial biosphere is assumed to be in equilibrium with the atmosphere, which is mixed on a global scale; the marine biosphere is assumed to be in equilibrium with the surface ocean. The deep ocean is neglected, since the time scale for mixing is too long (1000 years) for the period of consideration (100 years). Assume that alkalinity is due only to carbonates and that the input from rivers is balanced by removal of $CaCO_3$ in shallow-water sediments. A mass balance on the CO_2 added to the atmosphere (dG), which can be distributed either in the atmosphere (dP) or the ocean (dC_T), leads to (note G, P, and C_T must all be expressed in the same units):

$$\left(\frac{\partial G}{\partial P}\right)_A = 1 + \left(\frac{\partial C_T}{\partial P}\right)_A.$$

Following Eqs. (5.40) to (5.48) in the text, evaluate $(\partial G/\partial P)_A$.

Estimate the annual increase in partial pressure of CO_2 for an annual increase in CO_2 production of 4.2%, for a mixed ocean depth of 500 m. The present average atmospheric CO_2 content at sea level is $3.34 \cdot 10^{-4}$ atm (or 334 ppm by volume); the total mass of the atmosphere is $5.12 \cdot 10^{18}$ kg; the area of the ocean in $3.61 \cdot 10^{14}$ m^2; the density of seawater is approximately 1.023 kg/L. Compare your results with the observed increase of P at Mauna Loa observatory, Hawaii, from 315 ppm in 1958 to 331 ppm in 1974 (Fig. 5.11).

51. The partial pressure of CO_2 over a solution at constant A and C_T (as reflected by K_H) decreases with increasing temperature (Table 2.2 and Fig. 5.24). One concern about the influence of increased CO_2 on climate is that an increased average ocean temperature will diminish the capacity of the oceans to absorb CO_2. If the oceans are assumed to be fully mixed to 6000 m or more on the time scale of interest, their CO_2 content will be large compared to the atmosphere. If you further assume that the sediments do not react (since their reaction would decrease the possible magnitude of the effect), the appropriate coefficient is $(\partial P/\partial T)_{C_T,A}$. Evaluate this coefficient.

52. Making use of McIntyre's two-box model, according to which only the mixed layer (~ 500 m) of the ocean (and no sediment) is in equilibrium with the atmosphere, the appropriate coefficient is $(\partial P/\partial T)_{\Sigma,A}$, where Σ is the total of CO_2 in the atmosphere and in the mixed layer of the ocean ($G + C_T$ in Problem 50.)

53. In 1960, Lars Gunnar Sillén (an inorganic-physical chemist) attempted to explain how the composition of the oceans has remained essentially constant for about 100 million years.[*] He hypothesized that equilibria between water and a number of minerals could

---

[*] Sillén, L. G. 1961. In *Oceanography*, M. Sears, ed. American Association for the Advancement of Sciences Publ. No. 67; Sillén, L. G. 1967. *Science* 156:1189; Stumm, W. 1967. *Equilibrium Concepts in Natural Water Systems*. American Chemical Society for Advances in Chemistry Series No. 67; Stumm and Morgan, op. cit., pp. 572–574.

maintain both the known seawater composition and the partial pressure of CO_2 in the atmosphere. The simplest version of this model ($CaCO_3$, CO_2, and H_2O), as you have already seen, determines the pH, alkalinity, etc., when P_{CO_2} is specified. In general, the Gibbs phase rule gives the number of degrees of freedom as

$$F = C - P + 2,$$

where C is the number of chemical components (CaO, CO_2, and H_2O in the familiar example), P is the number of phases (atmosphere, water, and $CaCO_3$), and the factor 2 accounts for the possibility of external changes in temperature and pressure. If $F = 2$, temperature and pressure can be independently varied. This relation is derived in all physical chemistry textbooks from the fundamental principles of phase equilibria.

Addition of aluminosilicate minerals (for example, kaolinite and calcium montmorillonite) increases the number of components (adding Al_2O_3 and SiO_2) and phases to 5. You can keep the same two degrees of freedom, T and P, but now the dissolved silica content is fixed by equilibria between the phases. A simplified version (Stumm and Morgan, op. cit., pp. 530–531) is

$(CaO)(Al_2O_3)_7(SiO_2)_{22}(H_2O)_6 + 2H^+ + 23H_2O \rightleftharpoons$
calcium montmorillonite (s)

$$7(Al_2O_3)(SiO_2)_2(H_2O)_2 + 8Si(OH)_4(aq) + Ca^{++}$$
kaolinite (s)

with equilibrium constant (at 25°C)

$$K_{as} = \frac{[Si(OH)_4]^8[Ca^{++}]}{[H^+]^2} = 10^{-15.4}.$$

An additional aluminosilicate mineral (calcium feldspar) increases the number of phases but not the number of components, thereby decreasing the degrees of freedom to 1, which could be temperature. The solution equilibria combine to determine the partial pressure of CO_2 in the gas phase. Introducing an additional phase leads to an additional equilibrium, for example:

$(CaO)(Al_2O_3)(SiO_2)_2 + 2H^+ + H_2O \rightleftharpoons (Al_2O_3)(SiO_2)_2(H_2O)_2 + Ca^{++}$
calcium feldspar (s)   kaolinite (s)

$$K_f = \frac{[Ca^{++}]}{[H^+]^2} = 10^{+14.4}.$$

Combine these equilibria to predict P_{CO_2} for a model of the oceans consisting of CO_2, H_2O, $CaCO_3$, and the three aluminosilicate minerals above. (These simple and elegant equilibrium models have been abandoned because the necessary silicate phases are not found in ocean sediments in large enough quantities. The geochemists' emphasis has shifted to the balance between inputs and outputs.*)

54. Equal volumes of normal seawater (chlorinity 19.0‰) and a brackish water (chlorinity 2.0‰) are mixed. The brackish water is composed of normal seawater plus a river water containing alkalinity $0.4 \cdot 10^{-3}$ equivalents per liter, $[Ca^{++}] = 0.1 \cdot 10^{-3}$ M, and

---

* See Holland, op. cit.; McDuff, R. E. and Morel, F. M. M. 1980. *Environ. Sci. Technol.* 14:1182–1186.

$[Na^+] = 0.2 \cdot 10^{-3}$ M. Calculate the pH of the seawater, the brackish water, and the mixture, assuming that all are in equilibrium with atmospheric CO_2 at $10^{-3.5}$ atm.

55. These three solutions are each saturated with calcite at 10°C (see Table 5.8):

	Solution 1	Solution 2	Solution 3
pH	7.335	7.990	7.071
$[Ca^{++}]$	$2.12 \cdot 10^{-3}$	$0.964 \cdot 10^{-3}$	$14.64 \cdot 10^{-3}$
Alkalinity	$4.33 \cdot 10^{-3}$	$1.93 \cdot 10^{-3}$	$2.57 \cdot 10^{-3}$
P_{CO_2}	10^{-2} atm	10^{-3} atm	10^{-2} atm
$[SO_4^=]$			$13.35 \cdot 10^{-3}$

Show that the equimolar mixture of solutions 1 and 2 is undersaturated with calcite, but the equimolar mixture of solutions 1 and 3 is supersaturated with respect to calcite.

56. Big Soda Lake consists mostly of sodium carbonate, sulfate, and chloride (Truesdell, A. H. and Jones, B. F. 1969. *Chem. Geol.* 4:51–62). A full analysis of one sample gave (in mmole/L, except for pH):

$$[Na^+] \quad 374.5 \qquad [Mg^{++}] \quad 8.21 \qquad [SO_4^=] \quad 64.74$$
$$[K^+] \quad 0.99 \qquad C_T \quad 43.15 \qquad [Cl^-] \quad 213.5$$
$$[Ca^{++}] \quad 0.20 \qquad A \quad 65.54 \qquad pH \quad 9.6$$

Verify the charge balance (note whether the discrepancy is positive or negative) and decide whether it is reasonable in light of the major ion concentrations. Calculate the ionic strength without assuming any ion pairing and use that to estimate ion-pairing constants from the authors' (zero ionic strength) data. The number is the log of the formation constant (e.g., for the equilibrium $Na^+ + CO_3^= \rightleftharpoons NaCO_3^-$, $K = 10^{+1.27}$). Compare p. 125:

$NaCO_3^-$	$+1.27$	KSO_4^-	$+0.85$	$CaCO_3^0$	$+3.2$
$NaSO_4^-$	$+0.98$	$MgHCO_3^+$	$+1.00$	$CaSO_4^0$	$+2.3$
$NaHCO_3^0$	-0.25	$MgSO_4^0$	$+2.40$	$CaHCO_3^+$	$+1.26$
$Na_2CO_3^0$	$+0.68$	$MgCO_3^0$	$+3.4$		

Hint: You can set up a massive computer program to do these calculations, but a useful strategy (computer or not) is to make an initial computation of the concentration of each possible ion pair, assuming the constituent ions have the same concentration as their total analytical concentration. For example (since $[CO_3^=] \cong A - C_T$ and γ_+ is $\cong 10^{-0.15}$ at $I = 1$):

$$[NaCO_3] = 10^{+1.27}[Na^+][CO_3^=](\gamma_+)(\gamma_=)$$
$$\cong (10^{+1.27})(10^{-0.43})(10^{-1.65})(10^{-0.15})(10^{-0.60})$$
$$\cong 10^{-1.56}.$$

This is therefore an important species in the model, since it is of the same order of magnitude as $[CO_3^=]$, but its formation will not affect $[Na^+]$ very much. You can verify that other important species are $NaSO_4^-$, $MgSO_4^0$, and $MgCO_3^0$.

Complete the model, giving concentrations for free ionic species as well as ion pairs. Verify that the charge balance is the same as for the initial data. Check to see if any of the constants should be adjusted in the light of the revised ionic strength.

57. A strongly saline water (Field No. 39D, Deep Springs Lake, California; see Garrels and Christ op. cit., p. 109) had the following overall composition (in mole/L):

$[Na^+]$	6.48	$[HCO_3^-]$	0.30	pH	10.1
$[K^+]$	0.59	$[SO_4^=]$	0.62	Density	1.272 g/cm^3
$[CO_3^=]$	1.56	$[Cl^-]$	2.42		

Garrels and Christ calculated the activities of the major ions and ion pairs:

	Concentration	Activity		Concentration	Activity
Na^+	2.50	4.10	$CO_3^=$	0.20	0.02
$NaCO_3^-$	0.69	0.41	K^+	0.50	0.30
$NaSO_4^-$	0.37	0.22	KSO_4^-	0.09	0.054
$Na_2CO_3^0$	0.84	1.58	$SO_4^=$	0.15	0.016
$NaHCO_3^0$	0.20	0.38	Cl^-	2.42	1.43
HCO_3^-	0.10	0.06	I	4.54	

Verify these calculations. If the Deep Springs Lake brine were saturated with calcite, the solubility product ($10^{-8.52}$) of that mineral would give the activity of Ca^{++}. If the brine were also saturated with gypsum ($CaSO_4 \cdot 2H_2O$; $K_{s0}^0 = 10^{-4.37}$), the sulfate activity would determine the Ca^{++} activity. Test these hypotheses, knowing that the total Ca and Mg were too low to measure ($<10^{-5}$ M).

58. Calculate the activity of the ions and ion pairs that would result if the Deep Springs Lake brine (Problem 57) were diluted with pure water to one tenth of its original concentration.

59. If equal amounts of Deep Springs Lake brine (Problem 57) and Big Soda Lake water (Problem 56) were mixed, what would be the resulting concentration? Would the solubility product of calcite be exceeded?

60. A hydrothermal solution is saturated with calcite at 175°C, and initially completely fills a sealed rock cavity. As this rock cools to 20°C, the density of the solution increases from 0.892 to 0.998 g/cm^3 and a space containing water vapor and CO_2 appears (see Fig. 5.21). Calculate the concentration of dissolved carbonate species (including HCO_3^- and $CaHCO_3^+$) in the fluid. Find also the pressure of CO_2 in the vapor space and the ratio of total CO_2 in the vapor to that in the fluid. The solubility of CO_2 can be obtained from Fig. 5.24 and the solubility of $CaCO_3$ from Fig. 5.28, or you can interpolate the data in Table 2.2, and use Eq. (5.17). Notice that, because the pressure decreases along with the temperature, $CaCO_3$ might precipitate during cooling. Test for this possibility (see Ellis, A. J. 1963. *Am. J. Sci.* 261:259–267, especially p. 266).

61. To measure the acidity constants of CO_2 at high pressure, A. J. Ellis (1959. *Am. J. Sci.* 257:287–296) used a differential vapor pressure method, comparing the vapor pressure of a Na_2CO_3–$NaHCO_3$ solution with that of pure water. The difference was primarily due to the equilibrium partial pressure of CO_2, although the presence of the salts reduced

the water vapor pressure over the solution, and a correction had to be made for that. Here are some of Ellis's raw data:

Bomb volume	105.4 cm^3
Temperature	$162°C$
Na_2CO_3 added	0.01063 mole
$NaHCO_3$ added	0.0385 mole
Water	67.9 g
Vapor pressure of pure water at $162°C$	4877 Torr
Correction for salts (estimated)	116 Torr
Vapor pressure difference	129 Torr
Volume of solution at $162°C$	76.6 cm^3
Total CO_2 in system	
(calculated from P_{CO_2}, volumes, and K_H)	0.00048 mole
$NaHCO_3$ (corrected for loss to vapor)	0.03755 mole
Na_2CO_3 (corrected for loss to vapor)	0.01111 mole

Use these data to obtain a value for the combination $K_{a1}K_H/K_{a2}$ at $162°C$. What is the ionic strength of this solution?

62. Figures 5.27 and 5.28 show the solubility of calcite in water at temperatures from 100 to $300°C$ and at partial pressures of CO_2 from 1 to 62 atm. These data are related to the solubility product of calcite by the familiar Eq. (5.17). Strictly speaking, P_{CO_2} should be the fugacity, not the pressure, and the difference will be apparent as a slight systematic trend with increasing pressure; K_H^0 is given as a function of temperature in Fig. 5.24; K_{a1}^0 and K_{a2}^0 are given in Fig. 5.26. Estimate K_{s0}^0 for calcite at several temperatures from $100°C$ to $300°C$.

   Compare with Ellis (1963. *Am. J. Sci.* 261:259–267), who gives $pK_{s0} = 9.62$ at $100°C$, 10.54 at $150°C$, and 11.62 at $200°C$.

63. In the text (Fig. 5.31) you saw that the solubility of calcite at high carbonate concentrations (>1 M) increased as the third power of $[CO_3^=]$, whereas at low carbonate concentrations ($<10^{-5}$ M) the solubility product expression predicts that solubility of calcite should vary inversely with $[CO_3^=]$. There is little information about the intermediate range, but you will recall (from Eq. (4.8)) that the ion pair (or first complex) is given by

$$[CaCO_3^0] = K_{s1} = 10^{-5.42}$$

at $25°C$ in equilibrium with solid calcite. From Fig. 5.31, the constant

$$K_{s4} = \frac{[Ca(CO_3)_4^{6-}]}{[CO_3^=]^3}$$

can be estimated to be $10^{-3.9}$ at $200°C$, $10^{-3.3}$ at $300°C$, and $10^{-2.8}$ at $400°C$. Sketch a semiquantitative picture of calcite solubility at $25°C$ and $200°C$ over the range of $[CO_3^=]$ from 10^{-8} M to 10^{+1} M. Assume that K_{s1} varies from $10^{-7.6}$ at $200°$ to 10^{-11} at $400°C$, that K_{s0} is given by Table 2.2, and that K_{s4} is given by the intercept at $\log [CO_3^=] = 0$ on Fig. 5.31 and varies linearly with the reciprocal of absolute temperature. Set upper and lower bounds on the probable values for K_{s2} and K_{s3}.

Engineering Applications: Water Conditioning

There are many reasons to want to change the composition of a natural water source or a wastewater stream, and many of these involve changes in the carbonate equilibria. Most waters contain a significant quantity of dissolved carbon dioxide, and many contain significant amounts of calcium or magnesium. Engineers attempt to modify the composition of water to protect health or the environment, to minimize corrosion of equipment, or to optimize the water for some other chemical process.

In this chapter I will draw on your experience with Chapters 2 and 3 in discussing pH adjustment. When this adjustment is made with a strong acid (such as H_2SO_4) or a strong base (such as NaOH), the calculations are identical to those of the alkalinity titration. When CO_2 is added to reduce pH (*carbonation*) or removed to raise pH (*aeration*), the equations are the familiar ones for constant partial pressure from Chapter 2.

When lime (calcium oxide, CaO, or hydroxide, $Ca(OH)_2$) is used to raise pH, it behaves more or less as a strong base, except that sufficient lime will cause $CaCO_3$ to precipitate. The equations governing this were derived in Chapter 4. You can calculate whether a water is saturated with $CaCO_3$ or not by calculating the ion product $[Ca^{++}]$ $[CO_3^=]$ and seeing whether it exceeds K_{s0}, as in Chapter 5. An alternative test—the Langelier index—is presented here because of its widespread use in the engineering literature.

Scale formation, of course, is one manifestation of a supersaturated water; but a slight supersaturation tends to produce a coherent film that inhibits corrosion. Because seawater is normally supersaturated with $CaCO_3$, boilers using seawater are vulnerable to scale as well as corrosion. The addition of a judicious amount of CO_2 can create the optimum conditions; addition of a larger amount of CO_2 can be used to remove calcium carbonate scale that has become too voluminous.

Removal of Ca and Mg from municipal waters (*softening*) helps diminish boiler scale and makes the use of detergents (especially soap) more efficient. One of the classic methods (*lime-soda process*) makes use of $Ca(OH)_2$ to raise pH and supplements the natural alkalinity of the water with Na_2CO_3, so that maximum amounts of Ca and Mg are precipitated.

One example of a chemical engineering application is the treatment of reformer gas ($CO_2 + H_2$) with concentrated, high-temperature K_2CO_3 to remove CO_2, so that relatively pure hydrogen can be used in an alkaline fuel cell. Here the conditions are far from room temperature, but the principles are the same as those those governing the weathering of limestone by groundwater.

ESTIMATION OF IONIC STRENGTH

In Chapter 5, the compositions of rainwater, river water, groundwater, and seawater were known fairly completely, because the geochemists who analyzed them were interested in all the major ionic constituents. Such detailed analyses are expensive and time-consuming, however, and quite often an engineer receives an analysis containing only those quantities that are of critical importance, and a few other easily measured things.

To calculate the most accurate corrections for ionic strength (p. 30) you would want to know the total amount of monovalent ions, the total amount of divalent ions, etc. But without a complete analysis, this sort of breakdown is not available. What you will normally get is a number for total dissolved solids (TDS), which was long ago determined by evaporating the filtered water to dryness, but now is a transformed measurement of electrical conductivity,* much the same as *salinity* in Chapter 5 is actually a transformed electrical conductivity measurement. Indeed, the ASTM Standards for boiler makeup water include the following maximum values (in mg/L):

Dissolved solids	35000	Bicarbonate	600
Suspended solids	15000	Hardness (as $CaCO_3$)	5000
Chloride	19000	Alkalinity (as $CaCO_3$)	500
Sulfate	1400		

From the values for chloride, dissolved solids, and sulfate, you will recognize (compare Table 5.4) something like seawater, a little depleted in sulfate, but (with the suspended solids) very murky indeed. While this specification gives values for $[Cl^-]$, $[SO_4^=]$, $[HCO_3^-]$, and $[Ca^{++}] + [Mg^{++}]$, it does not give values for sodium, potassium, or borate, all of which are important for an oceanographer. Two of the items, bicarbonate and alkalinity, express the same quantity.

Faced with incomplete analyses, the engineer must make the best assumptions possible to estimate the ionic strength. You know that the ionic strength of seawater

---

* Lind, C. J. 1970. US Geol. Survey Prof. Paper 700D, pp. D272–280. His correlation between conductivity and ionic strength is reprinted by Snoeyink, V. L. and Jenkins, D. 1980. *Water Chemistry.* New York: Wiley, p. 77.

is 0.700 mole/kg or 0.716 mole/L, and one obvious assumption is that all waters otherwise unspecified have the same ratio of TDS to ionic strength as seawater

$$I = \frac{0.716 \text{ mole/L}}{35,000 \text{ mg/L}} (\text{TDS}) = 2.04 \cdot 10^{-5} (\text{TDS}), \tag{6.1}$$

where TDS is in mg/L. This is equivalent to assuming that the dissolved solids are mixtures of $1:1$ salts with an average formula weight of 48.9 g/mole. (That is lower than NaCl (58.5) because $[Mg^{++}]$ and $[SO_4^{=}]$ are multiplied by 4 in the ionic strength equation.)

Langelier, W. F. (1936. *J. Am. Water Works Assoc.* 28:1500–1521) used a conversion from TDS to ionic strength, based on Mississippi River water, that was equivalent to assuming a monovalent salt with formula weight 40 g/mole. Tables based on this conversion factor are in widespread use. In the absence of complete analytical data for the major ionic components of a water, I therefore recommend the assumption

$$I = \frac{1}{2} \sum c_i z_i^2 + 2.50 \cdot 10^{-5} R, \tag{6.2}$$

where R is residual TDS, the amount remaining when the amounts of all known components (in mg/L) have been subtracted from TDS: $R = \text{TDS} - \sum c_i M_i$, where M_i is the molecular weight of the species with concentration c_i in mole/L.

EXAMPLE 1   The composition of a water is found to be

Calcium hardness   182 ppm as $CaCO_3$ or $[Ca^{++}] = 1.82 \cdot 10^{-3}$ mole/L

Total hardness
(Mg and Ca)   205 ppm as $CaCO_3$ or $[Mg^{++}] = 2.30 \cdot 10^{-4}$ mole/L

Alkalinity   175 ppm as $CaCO_3$ or $A = 1.75 \cdot 10^{-3}$ mole/L

Total dissolved
solids   275 ppm by weight

pH   7.87

Temperature $50°F = 10°C$

Estimate the ionic strength. Note that the calcium ion, magnesium ion, and bicarbonate ion (alkalinity) make up

$$182 \frac{40}{100} + (205 - 182) \frac{24}{100} + 175 \frac{61}{100} = 185 \text{ mg/L},$$

and the remainder of the salts make up

$$275 - 185 = 90 \text{ mg/L}.$$

Without further analysis, the nature of these salts is unknown.*

The ionic strength from Eq. (6.1) is $5.6 \cdot 10^{-3}$, but it can be somewhat more precisely estimated from Eq. (6.2):

$$I = \frac{1}{2}([HCO_3^-] + 4[Ca^{++}] + 4[Mg^{++}] + \cdots) + (2.50 \cdot 10^{-5})(90),$$

$$= \frac{1}{2}[1.75 \cdot 10^{-3} + 4(1.82 \cdot 10^{-3} + 2.30 \cdot 10^{-4})] + 2.25 \cdot 10^{-3}$$

$$= 7.23 \cdot 10^{-3}, \tag{6.3}$$

where the three ellipsis dots represent negligible contributions from $CO_3^=$, H^+, OH^-, etc. Note that $A = [HCO_3^-]$, as is appropriate in this pH range (see Chapter 2), and that the divalent ions are multiplied by the square of their charge. Although this computation has been carried to three significant figures, the uncertainty in the nature of the unidentified anions and cations makes the R term uncertain by at least 20% ($\pm 5 \cdot 10^{-4}$), and hence the ionic strength is uncertain by about 10%. However, this is probably more accurate than either of the values based on TDS alone: $I = 5.6 \cdot 10^{-3}$ based on Eq. (6.1) and $I = 6.9 \cdot 10^{-5}$ based on 40 g/mole. What error is introduced to the activity coefficients because of the approximations in I? For $I = 7.2 \cdot 10^{-3}$, $t = 10°C$:

$$f(I)\left(\frac{I^{1/2}}{1 + I^{1/2}} - 0.2I\right)\left(\frac{298}{t + 273}\right)^{2/3} = 0.080. \tag{2.36) or (6.4}$$

For $I = 6.9 \cdot 10^{-3}$, $f(I) = 0.078$; and for $I = 5.6 \cdot 10^{-3}$, $f(I) = 0.071$. Thus a typical pH calculation (pp. 106–112 or 189–194) would be in error by at most one digit in the second decimal place. ■

pH ADJUSTMENT BY ADDITION OF ACID OR BASE

Even as a natural water may be titrated with a strong acid in the laboratory to yield a curve given by Eq. (3.6) and shown in Fig. 3.1, a municipal or process water can be treated with acid to change its pH. You might be tempted to take the acid needed as simply the difference in hydrogen ion concentrations of the

---

* In this sample, there is an excess of positive charge ($2[Ca^{++}] + 2[Mg^{++}] = 4.10 \cdot 10^{-3}$) over negative charge ($[HCO_3^-] = 1.75 \cdot 10^{-3}$) in the known constituents, and so the other ions (with net negative charge $2.35 \cdot 10^{-3}$) may be at least partly divalent (e.g., sulfate). This could mean, for example, that $[M^+] + 2[X^=] = 2.35 \cdot 10^{-3}$. But unless the ratio $[M^+]/[X^=]$ were known, the terms needed for the ionic strength, $[M^+] + 4[X^=]$, could not be evaluated.

initial and final solutions, but if you did, you would have forgotten all that you learned about titration curves and buffer capacity in Chapter 3. Indeed, because the most common pH range of interest for engineers is between 6 and 9, the calculations are simpler than they might otherwise be; a good first approximation (Figs. 2.1 and 2.4) is that $[HCO_3^-]$ is the dominant carbonate species for pH 7 to 9.5; at pH below 6.3, $[CO_2]$ becomes as large as $[HCO_3^-]$ or larger.

EXAMPLE 2   How much acid is required to change the pH of the water described in example 1 from 7.87 to 6.00? If the acid is added as concentrated H_2SO_4 or HCl, C_T can be assumed to remain unchanged* but A will be decreased by the equivalent amount of acid. Since the initial A is given to be $1.75 \cdot 10^{-3}$, and for this example dilution is unimportant, the simplest relationship for obtaining C_T is

$$C_T = \frac{A + G}{F}, \qquad \text{(2.30) or (6.5)}$$

where

$$F = \frac{K_{a1}[H^+] + 2K_{a1}K_{a2}}{[H^+]^2 + K_{a1}[H^+] + K_{a1}K_{a2}} \qquad \text{(3.7) or (6.6)}$$

and

$$G = [H^+] - \frac{K_w}{[H^+]}. \qquad \text{(3.8) or (6.7)}$$

(If you want to include dilution, use Eq. (3.6) instead of (2.30).)

As you saw in Fig. 2.4, $[HCO_3^-]$ is dominant at pH = 7.87 and both $[H^+]$ and $[OH^-]$ are very small. This translates into a first approximation where $F = 1$ and $G = 0$; hence

$$C_T = A + \cdots = 1.75 \times 10^{-3} \qquad \text{(6.8)}$$

You can expect that the second approximation, involving the activity coefficients and equilibrium constants, will not make much difference, and so the answer will be insensitive to those worrisome assumptions and calculations. To do them most efficiently, rewrite (6.6) as (3.11), or with the activity coefficients made explicit (see pp. 34–37):

$$F = \frac{K_{a1}^0 10^{-pH}(\gamma_0/\gamma_-) + 2K_{a1}^0 K_{a2}^0/\gamma_=}{10^{-2pH} + K_{a1}^0 10^{-pH}(\gamma_0/\gamma_-) + K_{a1}^0 K_{a2}^0/\gamma_=}. \qquad \text{(6.9)}$$

Recall Example 1 where you found $I = 7.2 \cdot 10^{-3}$ and $f(I) = 0.081$. This gives, via (2.36), $\gamma_- = 10^{-0.04}$ and $\gamma_= = 10^{-0.16}$. From Table 2.4, $b = 0.11$ at 10°C, and

---

* This assumption depends on three factors: C_T is not large (10^{-3}) compared to the strong acid concentration (>1 M), so there is negligible dilution; the final pH is not too low, so $[CO_2]$ is low and CO_2 is not lost to the atmosphere; and the temperature is relatively low (so CO_2 is more soluble and will be lost more slowly even if $[CO_2]$ exceeds saturation). This assumption would not be valid if $C_T = 10^{-1}$, pH = 4.0, for example, since then $P_{CO_2} > 1$ atm.

from Eq. (2.38) you find $\gamma_0 = 10^{-0.0007}$. From Table 2.2, you find $K^0_{a1} = 10^{-6.464}$ and $K^0_{a2} = 10^{-10.490}$ at 10°C. Putting all these values in (6.9), you obtain $F = 0.968$. Since $G = -7 \cdot 10^{-7}$, the second approximation is

$$C_T = \frac{1.75 \cdot 10^{-3} - 7 \cdot 10^{-7}}{0.968} = 1.81 \cdot 10^{-3}. \tag{6.10}$$

The complicated calculations of Eq. (6.9) changed the value of C_T by only 3%, and would not have been necessary for a rough calculation.

At the final pH = 6.0, you can calculate the alkalinity A_f by using (6.5) in the form

$$A_f = C_T F_f - G = C_T F_f + \cdots, \tag{6.11}$$

with F_f given by (6.6) or (6.9); G is given by (6.7) but is negligible at pH = 6. Because of the lower pH, the first term in the denominator of F is more important, but the last term in both numerator and denominator will be negligible (you can verify this). With the same values as above, except for pH, you can show that $F_f = 0.270$ and $A_f = 4.90 \cdot 10^{-4}$.

The amount of acid required to achieve this pH is

$$A - A_f = 1.26 \cdot 10^{-3}. \tag{6.12}$$

This could be supplied by H_2SO_4 (molecular weight 98.08) in the amount of

$$(1.26 \cdot 10^{-3} \text{ Eq/L})(49.04 \cdot 10^3 \text{ mg/Eq}) = 61.8 \text{ mg/L}.$$

Note finally that the amount of acid needed to change the pH from 7.87 to 6.00 is over 1000 times larger than the simple difference in hydrogen ion concentration:

$$\Delta[H^+] = 10^{-6.00} - 10^{-7.87} = 9.87 \cdot 10^{-7} \text{ Eq/L}. \qquad \blacksquare$$

As you may surmise, calculations for the adjustment of pH upward with strong base use the same equations. The only difference is that addition of one equivalent of base *increases* alkalinity by one equivalent.

ADDITION OR REMOVAL OF CO$_2$

As you saw in Chapter 2, addition or removal of CO_2 from a solution does not change the alkalinity, but can change the total dissolved carbonate and pH. Two important engineering processes are based on this: addition of CO_2 as a gas (*carbonation*) and removal of CO_2 by *aeration*. For drinking water (e.g., carbonated beverages), the CO_2 added must be pure, but for wastewaters or many industrial process waters a flue gas, consisting of about 20% CO_2 and 80% N_2 with traces of combustion products, is adequate.

EXAMPLE 3   What volume of flue gas per liter of water is required to change the pH of the water described in Example 1 from 7.87 to 6.00? The estimates of ionic strength and equilibrium constants, as well as many of the other calculations, were done in Example 2. We already know that $[HCO_3^-]$ is essentially equal to A, and this shortens a rough computation considerably.* In this case, A is held constant at $1.75 \cdot 10^{-3}$ and C_T is changed by addition of CO_2. In the initial solution, from Example 2, $C_T = 1.81 \cdot 10^{-3}$. In the final solution of pH = 6.00, the term corresponding to $[CO_3^=]$ can be neglected:

$$C_{Tf} = [HCO_3^-]\left(\frac{[H^+]}{K_{a1}} + 1 + \cdots\right)$$

$$= A\left(\frac{10^{-pH}}{K_{a1}^0(\gamma_0/\gamma_-)} + 1 + \cdots\right)$$

$$= (1.75 \cdot 10^{-3})(2.66 + 1.00 + \cdots) = 6.40 \cdot 10^{-3}.$$

The difference, $\Delta C_T = 4.60 \cdot 10^{-3}$ mole/L, is to be added from the flue gas. At 10°C, the Ideal Gas Law gives

$$V = \frac{nRT}{P} = \frac{(4.60 \cdot 10^{-3} \text{ mole})(0.08205 \text{ L·atm/mole·°K})(283°\text{K})}{(0.20 \text{ atm})}$$

$$= 0.533 \text{ liters of gas per liter of water.} \tag{6.13}$$

The fact that flue gas consists of only 20% CO_2 was taken into account by setting the partial pressure equal to 0.2 atm. Alternatively, the pressure could have been set equal to 1 atm, and the resultant volume multiplied by 1/0.2 to give the total volume of flue gas required. The answer would have been the same. Note that this is a far cheaper method than adding mineral acid if large quantities of water are to be treated.   ■

EXAMPLE 4   What is the lowest pH to which the water of the previous example could be brought by saturating it with flue gas? For this we require K_H from Table 2.2, at 10°C, $pK_H^0 = 1.27$. This should not change significantly at this low ionic strength (recall $\gamma_0 = 10^{-0.0007}$ in Example 2). As above, alkalinity is constant at $1.75 \cdot 10^{-3}$ and $P_{CO_2} = 0.20$ atm, so that when the water is saturated with flue

---

* If pH were higher, this approximation would not be so good, and you would have to use

$$A = [HCO_3^-] + 2[CO_3^=] + \cdots,$$

which gives

$$[HCO_3^-] = A\left(\frac{[H^+]}{[H^+] + 2K_{a2}}\right) = A\left(\frac{10^{-pH}}{10^{-pH} + 2K_{a2}^0(\gamma_-/\gamma_=)}\right) + \cdots.$$

For this problem, $[HCO_3^-] = 1.74 \cdot 10^{-3}$ at pH = 7.87.

gas, the pH is given by

$$A = [HCO_3^-] - [H^+] + \cdots = \frac{K_{a1}K_H P_{CO_2}}{[H^+]} - [H^+] + \cdots$$

$$= K_{a1}^0 K_H^0 P_{CO_2} \frac{10^{+pH}}{\gamma_-} - \frac{10^{-pH}}{\gamma_+} + \cdots. \tag{2.16}$$

Substituting numerical values, you will get

$$1.75 \cdot 10^{-3} = (0.2)(10^{-1.27-6.464+0.040+pH}) - 10^{-pH+0.040}.$$

This is a quadratic of the form

$$4.05 \cdot 10^{-9}(10^{+2pH}) - 1.75 \cdot 10^{-3}(10^{+pH}) - 1.10 = 0,$$

whose solution is pH = 5.64. Note that the last term is about 0.1% of the first two and that if the approximation $A = [HCO_3^-]$ had been used, the answer would have been the same.   ■

EXAMPLE 5   What is the highest pH to which the water of the previous examples could be brought by aeration? The answer to this question depends on the partial pressure of CO_2 in the aeration chamber, which will probably be somewhat higher than the mean atmospheric value of $10^{-3.5}$ atm. For this example, I assume $P_{CO_2} = 10^{-3}$ atm. Then the same equations as used above (neglecting $[H^+]$ this time) give

$$A = [HCO_3^-] + \cdots = \frac{K_{a1}^0 K_H^0 P \, 10^{+pH}}{\gamma_-}, \tag{6.14}$$

$$1.75 \cdot 10^{-3} = (10^{-3})(10^{-1.27-6.464+0.040+pH}),$$

$$pH = 7.94.$$

The approximations of neglecting $[H^+]$ and $[CO_3^=]$ are indeed justified (see Fig. 2.1).   ■

These examples have shown that, while carbonation and aeration are inexpensive, they have their limitations. To make pH lower than about 5.6 requires the addition of mineral acid, and to raise the pH above about 8 requires the addition of lime or sodium carbonate.

ADDITION OF LIME

$Ca(OH)_2$ increases both hardness (i.e., $[Ca^{++}]$) and alkalinity (by converting CO_2 to HCO_3^- or HCO_3^- to $CO_3^=$). If Q moles per liter of lime are added, $[Ca^{++}]$ increases by Q and A increases by $2Q$, but C_T remains invariant. The increase in pH for a given addition of lime can be calculated with the same equations as in the previous examples; small numerical differences may occur because addition of lime will tend to increase the ionic strength and change the activity coefficients. Large additions of lime, on the other hand, usually cause precipitation of $CaCO_3$ (which diminishes the ionic strength), and so in your calculations you should assume that the ionic strength is constant until after saturation with $CaCO_3$ has been investigated.

EXAMPLE 6   How much lime should be added to the water resulting from Example 2 (pH 6.00) to increase pH to 8.20? You can use some numerical results from Example 2:

$$C_{T0} = 1.81 \cdot 10^{-3}, \quad A_0 = 4.90 \cdot 10^{-4}, \quad pK_{a1}^0 = 6.464, \quad pK_{a2}^0 = 10.490,$$
$$I = 7 \cdot 10^{-3}, \quad \gamma_- = 10^{-0.04}, \quad \gamma_= = 10^{-0.16}, \quad \gamma_0 = 10^{-0.0007}.$$

The ionic strength and activity coefficients are provisional and will be reevaluated later.

To calculate the alkalinity A_f at pH = 8.20, use (6.5), (6.6), and (6.7), noting that $C_{Tf} = C_{T0}$, $G < 10^{-6}$, and $F_f \cong 1$:

$$A_f = C_{Tf}F_f + \cdots \cong 1.81 \cdot 10^{-3}.$$

More precisely, from (6.9), $F_f = 0.990$ and $A_f = 1.79 \cdot 10^{-3}$. The amount of alkali needed, therefore, is

$$A_f - A_0 = 1.79 \cdot 10^{-3} - 4.90 \cdot 10^{-4} = 1.30 \cdot 10^{-3} \text{ Eq/L.}$$

This can be supplied by $0.65 \cdot 10^{-3}$ mole/L of $Ca(OH)_2$, or 48.2 mg/L.   ∎

EXAMPLE 7   Is the water of Example 6, at pH = 8.20 after addition of $0.65 \cdot 10^{-3}$ mole/L of $Ca(OH)_2$, saturated with $CaCO_3$? The initial water (Example 1) contained 182 ppm Ca hardness, or $[Ca^{++}] = 1.82 \cdot 10^{-3}$ mole/L. To this we have added $0.65 \cdot 10^{-3}$ mole/L for a total of $[Ca^{++}]_f = 2.47 \cdot 10^{-3}$. The total carbonate $C_T = 1.81 \cdot 10^{-3}$ together with pH gives you the free carbonate ion concentration:

$$[CO_3^=] = \frac{C_T K_{a1}^0 K_{a2}^0 / \gamma_=}{10^{-2pH} + K_{a1}^0 \, 10^{-pH}(\gamma_0/\gamma_-) + K_{a1}^0 K_{a2}^0/\gamma_=}. \qquad \textbf{(2.21) or (6.15)}$$

If the solution is saturated,

$$[Ca^{++}][CO_3^=]\gamma_{++}\gamma_= \geq K_{s0}^0. \qquad \textbf{(2.50) or (6.16)}$$

Now, before putting numbers in (6.15) and (6.16), is a good time to reevaluate the ionic strength. Assume that negligible dilution has occurred since Example 1 (concentrations are in mmol/L):

	Example 1	Example 2	Example 6
$[Ca^{++}]$	1.82	1.82	2.47
$[Mg^{++}]$	0.23	0.23	0.23
$A \cong [HCO_3^-]$	1.75	0.49	1.81
Unknown $(2.5 \cdot 10^{-5} R)$	2.25	2.25	2.25
$[SO_4^=]$	0	0.63	0.63
pH	7.87	6.00	8.20
I (Eq. (6.2))	7.23	7.86*	9.82

* The added acid was assumed to be H_2SO_4; if it were HCl, this value would be 7.23, unchanged from Example 1.

Note that addition of sulfuric acid or addition of $Ca(OH)_2$ increases ionic strength. The revised activity coefficients (Eqs. (2.36) and (2.38)) are $\gamma_0 = 10^{+0.001}$, $\gamma_- = 10^{-0.046}$, $\gamma_= = 10^{-0.183}$. Substituting in (6.15) and noting that $1.81 \cdot 10^{-3} = 10^{-2.742}$, you get:

$$[CO_3^=] = \frac{10^{-6.464 - 10.490 + 0.183 - 2.742}}{10^{-16.400} + 10^{-6.464 - 8.20 + 0.001 + 0.046} + 10^{-16.778}}$$

$$= 10^{-4.906}.$$

Now use (6.16) to test if the solubility product is exceeded. Recall that $[Ca^{++}] = 2.47 \cdot 10^{-3} = 10^{-2.607}$; then

$$[Ca^{++}][CO_3^=]\gamma_{++}\gamma_= = 10^{-2.607 - 4.906 - 2(0.183)} = 10^{-7.879}.$$

The solubility product of calcite at 10°C (Table 2.2) is $K_{s0}^0 = 10^{-8.41}$ (if ion pairs are included in the model). Thus the solubility product is exceeded by $10^{+0.53}$ (about a factor of 3) and you would expect $CaCO_3$ to precipitate only under favorable conditions. Recall from pp. 135–141 and the associated examples that surface ocean water is supersaturated with calcite by a factor of 2 to 6 but only rarely precipitates without biological assistance. ■

ADDITION OF LIMESTONE

If a water contains excess acidity, limestone $(CaCO_3)$ is a cheaper way of raising the pH than lime $Ca(OH)_2$, but as you will recall from the discussions of river and groundwaters in Chapter 5, it has its limitations, tending to bring the pH to around 8.3 while simultaneously increasing calcium ion concentration and alkalinity. The equations that govern a saturated $CaCO_3$ solution were developed in Chapter 4, and elaborated in Chapter 5, so that their application to water conditioning should be straightforward, even though the equations are more complicated than in the previous examples of this chapter. (To make the equations look more like those in Chapter 4, and also as a variation on handling activity coefficients, the following example will use concentration constants and pH $= -\log[H^+]\gamma_+$).

EXAMPLE 8   Water at 25°C with alkalinity $1.0 \cdot 10^{-3}$, $[Ca^{++}] = 8.0 \cdot 10^{-4}$ and pH $= 7.5$ is mixed with excess $CaCO_3$. What is the alkalinity and pH when equilibrium is reached? As you may recall from Chapter 4, this alkalinity is approximately equal to that obtained from $CaCO_3$ in pure H_2O, with CO_2 at atmospheric partial pressure, but the pH is a little lower and $[Ca^{++}] > \frac{1}{2}A$.

Is the initial solution saturated with $CaCO_3$ already? Test this by calculating the ion product (assume* $[HCO_3^-]_0 = A_0 = 10^{-3}$, $K_{a2} = 10^{-10.24}$, $[H^+]_0 = 10^{-7.48}$):

$$[Ca^{++}]_0[CO_3^=]_0 = \frac{[Ca^{++}]_0 K_{a2}[HCO_3^-]_0}{[H^+]_0} = 10^{-8.86}.$$

Since this result is less than K_{s0} ($10^{-8.34}$), the solution is undersaturated. A more precise calculation would use Eqs. (6.15) and (6.16) of Example 7, but would not change the conclusion. Since D is invariant to the addition of $CaCO_3$, you can write two relationships between the initial (subscript zero) and final solutions:

$$D_1 = \frac{1}{2}D = C_{T0} - [Ca^{++}]_0 = C_T - [Ca^{++}], \qquad \text{(4.40) or (6.17)}$$

$$D_2 = D - [Na^+] + [Cl^-] = [HCO_3^-]_0 + 2[CO_2]_0 + \cdots$$
$$= [HCO_3^-] + 2[CO_2] + \cdots. \qquad \text{(4.42) or (6.18)}$$

With the knowledge that pH will increase from 7.5, and will probably be between 8 and 9 in the final solution, I have neglected $[CO_2]$, $[H^+]$, $[OH^-]$, and the ion pairs in (4.42). Now use the appropriate equilibria[†]

$$A = [HCO_3^-]\left(1 + \frac{2K_{a2}}{[H^+]}\right) + \cdots, \qquad \text{(2.16) or (6.19)}$$

---

* If $I = \frac{1}{2}(A + 4[Ca^{++}] + [M^+]) = 2.2 \cdot 10^{-3}$, then $\gamma_\pm = 10^{-0.022}$, $\gamma_0 = 1$, $\gamma_= = 10^{-0.089}$. Then $K_H = K_H^0 = 10^{-1.47}$, $K_{a1} = K_{a1}^0 \gamma_0/\gamma_+\gamma_- = 10^{-6.308}$, $K_{a2} = K_{a2}^0(\gamma_-/\gamma_+\gamma_=) = 10^{-10.240}$, $K_{s0} = K_{s0}^0/\gamma_{++}\gamma_= = 10^{-8.34}$, $[H^+] = 10^{-pH+0.022}$.

† Note that I have expressed A and C_T separately in terms of $[HCO_3^-]$ and pH instead of using (2.30) or (6.5) and calculating F from pH (as in Examples 2 and 6) because I need $[HCO_3^-]$ and $[CO_2]$ for Eq. (4.42). Other combinations would work as well.

$$C_T = [HCO_3^-]\left(1 + \frac{K_{a2}}{[H^+]} + \frac{[H^+]}{K_{a1}}\right), \qquad \text{(2.22) or (6.20)}$$

$$[CO_2] = [HCO_3^-]\frac{[H^+]}{K_{a1}}, \qquad \text{(2.2) or (6.21)}$$

to obtain the functions of Eqs. (6.17) and (6.18):

$$[HCO_3]_0 = \frac{1.0 \cdot 10^{-3}}{1 + 10^{+0.30 - 10.24 + 7.48}} = 9.97 \cdot 10^{-4},$$

$$C_{T0} = (9.97 \cdot 10^{-4})(1 + 10^{-10.24 + 7.48} + 10^{-7.48 + 6.31}) = 1.07 \cdot 10^{-3},$$

$$[CO_2]_0 = (9.97 \cdot 10^{-4})(10^{-7.48 + 6.31}) = 6.74 \cdot 10^{-5},$$

$$D_1 = \frac{1}{2}D = 1.07 \cdot 10^{-3} - 8.0 \cdot 10^{-4} = 2.07 \cdot 10^{-4},$$

$$D_2 = 9.97 \cdot 10^{-4} + (2)(6.74 \cdot 10^{-5}) = 1.13 \cdot 10^{-3}.$$

Now, because D_1 and D_2 are invariant with respect to addition or removal of $CaCO_3$, they will have the same values in the final saturated solution:

$$D_1 = C_{Tf} - [Ca^{++}]_f = 2.07 \cdot 10^{-4}, \qquad \text{(6.22)}$$

$$D_2 = [HCO_3]_f + 2[CO_2]_f = 1.13 \cdot 10^{-3}. \qquad \text{(6.23)}$$

Here C_T, $[HCO_3^-]$, and $[CO_2]$ are related to pH by (6.20) and (6.21) above, and the final equation needed to solve (6.22) and (6.23) is provided by the solubility product:

$$[Ca^{++}][CO_3^=] = \frac{[Ca^{++}][HCO_3^-]K_{a2}}{[H^+]} = K_{s0}. \qquad \text{(6.24)}$$

Then from (6.20), (6.24), and (6.22), you get

$$[HCO_3^-]_f\left(1 + \frac{K_{a2}}{[H^+]_f} + \frac{[H^+]_f}{K_{a1}}\right) - \frac{K_{s0}[H^+]_f}{[HCO_3^-]_f K_{a2}} = 2.07 \cdot 10^{-4}. \qquad \text{(6.25)}$$

From (6.21) and (6.23), you get

$$[HCO_3^-]_f\left(1 + \frac{2[H^+]_f}{K_{a1}}\right) = 1.13 \cdot 10^{-3}. \qquad \text{(6.26)}$$

Either $[HCO_3^-]$ or $[H^+]$ could be eliminated between (6.25) and (6.26) to give one giant polynomial, but a simplification is obvious in the second term of (6.26) because the pH will be around 8:

$$[HCO_3^-]_f = \frac{1.13 \cdot 10^{-3}}{1 + 10^{+0.3 - 8.0 + 6.31}} \simeq 1.18 \cdot 10^{-3}.$$

Similarly, in (6.25) you can expect the second and third terms multiplying $[HCO_3^-]$ to be small, and you can estimate them by setting $[H^+] = 10^{-8}$ (don't set $[H^+] = 10^{-8}$ everywhere, however, or you'll lose your answer and end up with an unbalanced equation!):

$$(1.18 \cdot 10^{-3})(1 + 10^{-10.24+8.00} + 10^{-8.00+6.31}) - \frac{10^{-8.34}[H^+]_f}{(1.18 \cdot 10^{-3})(10^{-10.24})}$$

$$= 2.07 \cdot 10^{-4}.$$

Thus, $[H^+]_f = 10^{-7.83}$ and pH = 7.85. The alkalinity is given by (6.19):

$$A = 1.19 \cdot 10^{-3}. \quad \blacksquare$$

In general, if you make the approximation pH = 8 in the presence of solid calcium carbonate, you will not be far off (but see problem 9 for an exception!) and will more easily recognize when the terms you think were negligible are important.

If you would like a simple algebraic form for the approximate answer to a problem of this type, note that neglecting the second and third terms of (6.25) and the second term of (6.26) leads to

$$[HCO_3^-] - \frac{K_{s0}[H^+]}{[HCO_3^-]K_{a2}} = D_1 + \cdots,$$

$$[HCO_3^-] = D_2 + \cdots.$$

When these two equations are combined to eliminate $[HCO_3^-]$ and you set $D_1 = D/2$, the result is

$$[H^+] = \frac{K_{a2}}{K_{s0}} D_2(D_2 - D_1) + \cdots = \frac{K_{a2}}{2K_{s0}} D_2(2D_2 - D) + \cdots, \quad \textbf{(4.47) or (6.27)}$$

which was derived in Chapter 4 as Eq. (4.47). Do not forget about the limited pH range over which this applies. For example, $CaCO_3$ and H_2O with no added acid or base would give $D_2 = D = 0$ and pH = ∞. As you know, the pH for that solution is only about 9.95. For Example 8, however, (6.27) gives a fairly precise answer:

$$[H^+] = \frac{10^{-10.24}}{10^{-8.34}} (1.13 \cdot 10^{-3})(1.13 \cdot 10^{-3} - 2.07 \cdot 10^{-4}) = 10^{-7.88}.$$

THE LANGELIER INDEX OF CaCO₃ SATURATION

One measure of supersaturation has already been used in Examples 7 and 8 above: the ratio of the ion product $[Ca^{++}][CO_3^=]$ to the concentration solubility product K_{s0} of calcite.

A more popular measure of supersaturation in the older engineering literature is the *Langelier index*,* defined as the difference between the observed pH and the hypothetical pH at which the solution would be saturated with calcium carbonate:

$$L = pH_{obs} - pH_{sat}. \qquad (6.28)$$

(L is also designated as *S.I.*, *L.I.*, *I*, etc. in various texts and monographs.) A negative index indicates undersaturation, a positive index indicates oversaturation. Since $[Ca^{++}]$ and $[CO_3^=]$ are linked by the solubility product in a saturated solution, pH_{sat} can be expressed as a function of $[Ca^{++}]$ (hardness) and alkalinity alone if certain approximations are met. Here is the rigorous derivation:

$$A = [HCO_3^-] + 2[CO_3^=] + [OH^-] - [H^+]. \qquad (2.14)$$

Make use of the hybrid pH–concentration constants K_w', K_1', and K_2' as defined by Eqs. (2.47), (2.39), and (2.42), to obtain the carbonate ion concentration as a general function of A, pH, and ionic strength (recall that $\log \gamma_+ = -0.5f(I)$, where $f(I)$ is given by Eq. (2.36)):

$$[CO_3^=] = K_2' 10^{+pH} \frac{A - 10^{-pK_w' + pH} + 10^{-pH + 0.5f(I)}}{1 + 2K_2' 10^{+pH}}. \qquad (6.29)$$

If the solution is saturated,

$$[CO_3]_s [Ca^{++}]_s = K_{s0}. \qquad \textbf{(4.1) or (6.30)}$$

Combine this with (6.29) to get

$$[Ca^{++}]K_2' 10^{+pH_s}(A - 10^{-pK_w' + pH_s} + 10^{-pH_s + 0.5f(I)}) = K_{s0}'(1 + 2K_2' \cdot 10^{+pH_s}).$$

Take the logarithm and rearrange to get

$$pH_s = pK_2' - pK_{s0} - \log[Ca^{++}] + \log(1 + 10^{+0.30 - pK_2' + pH_s})$$
$$- \log(A - 10^{-pK_w' + pH_s} + 10^{-pH_s + 0.05f(I)}). \qquad (6.31)$$

Since the last two terms have only a slight pH dependence in the range from 6 to 9 (Fig. 6.1), this equation can be simplified to Langelier's formula

$$L = pH_{obs} - pH_s = pH_{obs} - (pK_2' - pK_{s0}) + \log A + \log[Ca^{++}] + \cdots. \qquad (6.32)$$

You have already seen a similar equation, (4.57). For solutions approaching zero ionic strength at $25°C$, $pK_2' = 10.3$ and $pK_{s0} = 8.3$ (Table 2.1),[†] so that their differ-

---

* Langelier, W. F., 1936. *J. Am. Water Works Assoc.* 28:1500–1521; Larson, T. E. and Buswell, A. M. 1942. *J. Am. Water Works Assoc.* 34:1667; Lowenthal, R. E. and Marais, G. vR. 1976. *Carbonate Chemistry of Aquatic Systems.* Ann Arbor, Mich: Ann Arbor Science Publishers, pp. 128–136.

[†] If the alternative set of solubility products in Table 2.2 based on the ion-pairing model is used, you find $pK_{a2}^0 - pK_{s0}^0 = 1.81$. At high CO_2 partial pressures, hardness includes $[CaHCO_3^+]$ as well as $[Ca^{++}]$. This reduces the simplicity of Langelier's formulation. For critical work, however, the ion-pairing model is desirable (see Problem 12).

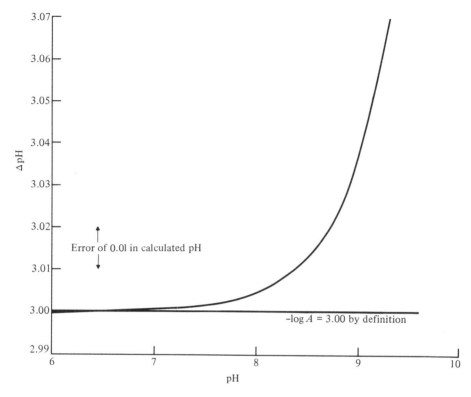

Fig. 6.1.   Terms neglected in the simplification of Eq. (6.31) to obtain Langelier's equation (6.32). At 25°C, with $I = 1.5 \cdot 10^{-3}$, $pK_2' = 10.274$, $pK_w' = 13.981$, $A = 10^{-3}$, the function plotted is

$$\Delta pH_s = \log(1 + 10^{-9.973 + pH_s}) - \log(A - 10^{-13.981 + pH_s} + 10^{-pH_s + 0.018}).$$

Note that the major deviation comes from the first term and corresponds to neglecting $[CO_3^=]$ compared to $[HCO_3^-]$ in the expression for alkalinity.

ence (Langelier's constant) is approximately 2.0. This result is consistent with the discussion of a solution saturated with CaCO₃ in Chapter 4 (see p. 83): if $A = 10^{-3}$ and $[Ca^{++}] = 0.5 \cdot 10^{-3}$ (e.g., Fig. 4.5) and $L = 0$, Eq. (6.32) gives $pH_{obs} = 8.3$.

At higher ionic strengths, the hybrid equilibrium constants both change, and Langelier's constant increases. Using the Davies equation to predict activity coefficients, you can show from Eqs. (2.43) and (2.51) that

$$pK_2' - pK_{s0} = pK_{a2}^0 - pK_{s0}^0 + 2.5f(I), \tag{6.33}$$

where $f(I)$ is given by Eq. (2.36). This difference has the shape shown in Fig. 6.2. Note that experimental values for seawater agree well with the Davies equation,

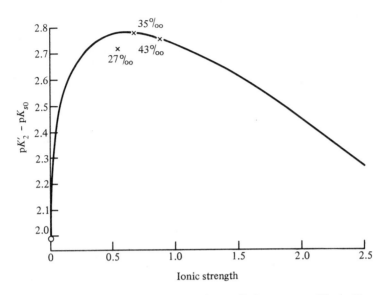

Fig. 6.2. Ionic-strength dependence of Langelier's constant $pK'_2 - pK_{s0}$ at 25°C. The curve is given by the Davies equation ((2.36) or (6.33)). The point for 35‰ seawater was taken from Table 5.2 and represents measurements by a number of workers; the points for 27‰ and 43‰ seawater were obtained by Ingle and Pytkowicz (see Fig. 5.9 and Ingle, S. E. 1975. *Mar. Chem.* 3:301–319). At zero ionic strength, pK^0_{s0} was taken to be 8.34, to agree with the older literature. If $pK^0_{s0} = 8.52$, the intercept would be 1.81 instead of 1.99. See Table 2.2 and the beginning of Chapter 4 for a discussion of the influence of ion pairs on the measurement of the solubility product for calcite.

in contrast to the poor agreement obtained for the individual constants in Figs. 2.7 and 2.9. This is because the activity coefficient of $CO_3^=$ cancels out, and that is the factor most affected by ion pairing. Values at other temperatures are shown in Figs. 6.3 and 6.4. Langelier, in his 1936 paper (op. cit.) used $f(I) = I^{1/2}$ as well as data for K^0_{s0} at temperatures from 0 to 100°C that are now obsolete. He revised those tables for a nomograph (C.P. Hoover (1938). *J. Am. Water Works Assoc.* 30:1802) and published them in full as a part of the discussion of Larson and Buswell's 1942 paper (op. cit.).

In 1942 Larson and Buswell (op. cit.) published an equation giving the Langelier constant as a function of total dissolved solids based on new thermodynamic data. Their conversion between ionic strength and total dissolved solids was made as if the electrolyte were univalent with a formula weight of 40 g/mole (see Eq. (6.2) and Example 1), and the activity coefficients were given by a function

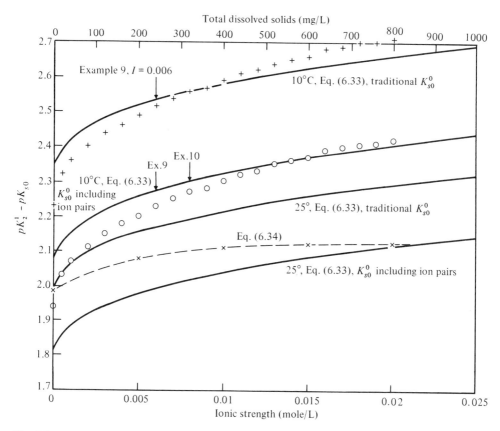

Fig. 6.3. Comparison of values for Langelier's constant $pK_2' - pK_{s0}$ compiled by Langelier (1942, in a lengthy comment to Larson and Buswell (op. cit.) pp. 1679–84), and frequently republished in the engineering literature, with the Davies equation at low ionic strength: $I < 0.025$ (corresponding to total dissolved solids up to 1000 mg/L). (+) Langelier's table, 10°C; (⊙) Langelier's table, 25°C; solid line—Davies equation (2.36); broken line—ionic strength function (Eq. (6.34)) used by Larson and Buswell (op. cit.)

that was less dependent on ionic strength than the Davies equation:

$$f(I) = \frac{I^{1/2}}{1 + 5.3I^{1/2} + 5.5I}. \tag{6.34}$$

This equation is an early empirical form that was superseded by the Guggenheim form, of which the Davies equation is an example.

Larger than the discrepancy resulting from the use of Eq. (6.34) to estimate activity coefficients is the discrepancy resulting from the use of traditional values

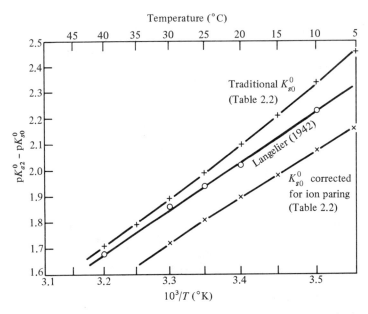

Fig. 6.4.   Temperature dependence of Langelier's constant at zero ionic strength: $pK_{a2}^0 - pK_{s0}^0$. Data from Table 2.2 for the traditional solubility product of calcite (+) and for the solubility product obtained with the ion-pairing model of Christ, C. L., Hostetler, P. B., and Siebert, R. M. (1974. *J. Res. US Geol. Survey* 2:175–184) (×) are compared with the values published by Langelier (op. cit.) (⊙).

of the solubility product instead of the more recently recalculated values (unknown, of course, in 1942) that take account of ion pairs (Table 2.2 and Chapter 4). This is obvious from the comparison made in Figs. 6.3 and 6.4. The activity coefficient discrepancy results in errors of less than 0.1 logarithmic units in L (or predicted pH), and the different set of equilibrium constants shifts the values by 0.2 to 0.3 logarithmic unit. Nevertheless, the Langelier tables (often attributed to Larson and Buswell) are still being quoted in engineering tables and textbooks to two decimal places.*

EXAMPLE 9   Is the water of pH = 7.87, described in Example 1, saturated with $CaCO_3$? The calculation can be done quickly with Langelier's formula. The constant $pK_2' - pK_{s0}$ obtained from Fig. 6.3 at 10°C and $I = 0.006$ is approximately 2.53 according to Langelier, and 2.26 according to Eq. 6.33 and

---

* See, for example, Clark, J. W., Viessman, W., and Hammer, M. J. 1977. *Water Supply and Pollution Control*, 3rd ed. New York: Harper & Row, p. 441.

the constants (ion-pairing model) from Table 2.2. The values from Example 1: pH $= 7.87$, $[Ca^{++}] = 1.82 \times 10^{-3}$, and $A = 1.75 \times 10^{-3}$, substituted in (6.32), give $L = -0.16$ (undersaturated) from the first estimate of $pK_2' - pK_{s0}$ and $+0.11$ (supersaturated) from the second.   ■

CORROSION INHIBITION

Waters that are highly buffered and slightly oversaturated with calcium carbonate tend to be less corrosive than poorly buffered or undersaturated waters. A coherent film of $CaCO_3$ helps protect the metal surfaces inside of pipes. Waters that are poorly buffered will fluctuate between oversaturation and undersaturation, and the calcium carbonate deposit will be *tubercular*, loosely adherent, and offer little corrosion protection.*

Adjustment of waters to achieve slight oversaturation as well as high buffer capacity is thus an important part of water conditioning.

EXAMPLE 10   For corrosion control, it is desirable to have the saturation index L slightly positive. How much $Ca(OH)_2$ must be added to the water described in Example 2 (pH $= 6.00$) so that $L = +0.04$? Recall that the alkalinity was $A_0 = 4.90 \times 10^{-4}$, the original Ca^{++} concentration was $1.82 \cdot 10^{-3}$, and the ionic strength was $7.9 \cdot 10^{-3}$. After the addition of x moles per liter of $Ca(OH)_2$, you will have (compare with Example 6)

$$A_f = A_0 + 2x, \qquad [Ca^{++}]_f = [Ca^{++}]_0 + x.$$

Langelier's equation (6.32) gives a relationship between pH and x:

$$L = +0.04 = pH_f - (pK_2' - pK_{s0}) + \log(A_0 + 2x) + \log([Ca^{++}] + x).$$

As in Example 9, you can get your data from Eq. 6.33 and Table 2.2 or Fig. 6.3 for these conditions ($10°C$ and $I = 0.0079$: $pK_2' - pK_{s0} = 2.30$). But until you have a value for x, or another relationship between pH and x, the problem is not yet solved. If you neglect x in the equation for L, you obtain $pH_f = 8.37$, which sounds like a good value. The hard part is to calculate a value for x, so you will know how much lime to add. This is done (as in Example 6) via the alkalinity equation

$$A = C_T F + G, \tag{2.29}$$

and with A approximately 10^{-3} at pH $= 8$, G (which is $[H^+] - [OH^-] = -10^{-6}$) is negligible; F is the usual function of pH ((3.7) or (3.9)). The key to these calculations is that C_T does not change when $Ca(OH)_2$ is added, even though A increases and so does pH. Therefore, you can calculate C_T from the initial conditions

---

* Stumm, W. 1960. *ASCE Journal, Sanitary Eng. Div.* 86 (Nov.) 27–45. Paper No. 2657.

and it will be the same for the final conditions:

$$C_T = \frac{A_0}{F_0} + \cdots, \tag{2.30}$$

$$A_f = A_0 + 2x = C_T F_f = \frac{A_0 F_f}{F_0} + \cdots,$$

$$x = \frac{A_0}{2}\left(\frac{F_f}{F_0} - 1\right) + \cdots. \tag{6.35}$$

Now all you need to do is evaluate F_f and F_0 at pH 8.37 and 6.00, respectively, using a plot like Fig. 3.2 constructed for 10°C and $I = 0.0079$, or Eq. (3.7). Be careful not to drop too many terms (as I have often suggested you should do), because some of the large terms cancel each other and the small terms carry the information you want. Without approximations, you get

$$\left(\frac{F_f}{F_0} - 1\right) =$$

$$K_{a1}\left[\frac{\begin{array}{c}K_{a2}[H^+]_0(K_{a1} + 2[H^+]_0) - K_{a2}[H^+]_f(K_{a1} + 2[H^+]_f) \\ + [H^+]_0[H^+]_f([H^+]_0 - [H^+]_f)\end{array}}{([H^+]_0^2 + K_{a1}[H^+]_0 + K_{a1}K_{a2})([H^+]_f^2 + K_{a1}[H^+]_f + K_{a1}K_{a2})}\right]. \tag{6.36}$$

In the pH range around 8, *all* the terms in (6.36) are the same within two orders of magnitude; the first and third terms in each of the denominator factors are about 10^{-2} of the middle term, but become more important at higher and lower pH. If you drop all but the largest terms, you can get a very approximate version applying near pH $= 8$:

$$\frac{F_f}{F_0} - 1 = ([H^+]_0 - [H^+]_f)\left(\frac{K_{a2}}{[H^+]_0[H^+]_f} + \frac{1}{K_{a1}}\right) + \cdots. \tag{6.37}$$

For this example, the full Eq. (6.36), with $[H^+]_0 = 10^{-5.96}$, $[H^+]_f = 10^{-8.33}$, $K_{a1} = 10^{-6.38}$, $K_{a2} = 10^{-10.33}$ gives $10^{-0.14}$. (Note the approximate Eq. (6.37) gives $10^{+0.42}$.)

Finally, use (6.35) to obtain

$$x = \frac{4.90 \times 10^{-4}}{2}(10^{-0.14}) = 1.77 \times 10^{-4} \text{ mole/L}.$$

This would be 13 mg/L of $Ca(OH)_2$. A second iteration with this new value of x gives pH $= 8.11$, and with this revision, Eq. (6.36) gives $x = 1.73 \times 10^{-4}$. A third iteration gives pH $= 8.12$. ■

EXAMPLE 11 A desalination plant experiences problems with excessive calcium carbonate scale in its boiler tubes and plans to add CO_2 from flue gas to the feed water. What volume of flue gas (20% CO_2, 80% N_2) is required per liter of seawater to reduce the Langelier index to $+0.10$ at $100°C$? The exact equilibrium constants needed are not found in Table 2.2, but the difference $pK'_2 - pK_{so}$ is shown on Fig. 6.2 as approximately 2.78 in 35‰ seawater at $25°C$. Note that the experimental data agree well with the Davies equation prediction (Eq. (6.33)) and that $I = 0.7$ is near the maximum of the curve. Therefore changes in concentration of seawater during distillation would not be expected to change the equilibrium constants very much. To get the value at $100°C$, use data from Table 2.2 for zero ionic strength and make use of (6.33) with $I = 0.7$.

Note that $f(I) = 0.272$ and $pK'_2 - pK'_{so} = 1.22$. This is not very accurate, since the Davies equation is meant to be used only near $25°C$ and in any case does not account for changes in ion pairing with temperature. The conversion from moles/L to moles/kg is negligible (1.23 instead of 1.22) compared to these other uncertainties. However, the alternative is to extrapolate experimental seawater data from $30°$ to $100°C$, which is probably still less accurate.

Addition of CO_2 changes neither alkalinity nor $[Ca^{++}]$, so that these can be obtained from Table 5.4. Note that, before addition of CO_2, at $pH_0 = 8.0$, seawater is highly oversaturated. From the Langelier equation derived above:

$$L = 8.0 - 1.22 + \log(2.3 \cdot 10^{-3}) + \log(10.33 \cdot 10^{-3}) = +2.15. \qquad (6.32)$$

To reduce L to $+0.10$, enough CO_2 must be added to reduce pH to

$$pH_f = +0.10 + 1.22 - \log(2.3 \cdot 10^{-3}) - \log(10.33 \cdot 10^{-3}) = 5.94.$$

Since A is constant, the difference in C_T can be calculated from the familiar definition of alkalinity (2.28), or (2.30) combined with (3.11). At $25°C$ and $pH = 8.0$, with $pK'_1 = 6.00$ and $pK'_2 = 9.12$ (Table 5.5):

$$C_{T0} = \frac{A}{F_0} + \cdots = 2.3 \cdot 10^{-3} \frac{10^{-pH-pK'_1} + 10^{-pK'_1-pK'_2} + 10^{-2pH'}}{10^{-pH-pK'_1} + 10^{+0.30-pK'_1-pK'_2}} + \cdots,$$

$$C_{T0} = 2.15 \cdot 10^{-3}.$$

(Note that $F_0 = 1.070$.) After adding CO_2 to reduce pH to 5.89 at $100°C$, calculate the total carbonate by the same equation

$$C_{Tf} = \frac{A}{F_f} + \cdots, \qquad (2.30)$$

but now use estimates of pK'_1 and pK'_2 at $100°C$. You may be tempted to use the Davies equation (i.e., (2.40) and (2.43)) to calculate the constants from zero ionic strength data in Table 2.2 (you'll get $pK'_1 = 6.25$ and $pK'_2 = 9.78$), but don't do it. Remember that the acidity constants in seawater (Figs. 2.6 and 2.9) are very much lower than predicted by the Davies equation because of ion pairing. It is only a fortuitous cancellation of the ion-pairing effects on $CO_3^=$ that allowed us to use the Davies equation to predict $pK'_2 - pK'_{so}$.

How should these constants be estimated then? One assumption is that the temperature dependence of the seawater constants is the same as the temperature dependence of the zero ionic strength constants. This is not too good, but it is better than the alternatives. From Tables 5.5 and 2.2, you can estimate

$$pK_1'(100°, SW) = pK_1'(25°, SW) + pK_{a1}^0(100°) - pK_{a1}^0(25°)$$
$$= 6.00 + 6.45 - 6.35 = 6.10$$
$$pK_1'(100°, SW) = 9.12 + 10.16 - 10.33 = 8.95.$$

Then from Eq. (3.11) you get

$$F_f = \frac{10^{-5.89-6.10} + 10^{+0.30-6.10-8.95}}{10^{-2(5.89)} + 10^{-11.99} + 10^{-15.05}} = 0.382$$

and from (2.30)

$$C_{Tf} = \frac{2.3 \cdot 10^{-3}}{0.382} = 6.02 \cdot 10^{-3}.$$

Thus the amount of CO_2 that needs to be added is

$$C_{Tf} - C_{T0} = 3.87 \cdot 10^{-3} \text{ mole/L}.$$

If the flue gas is measured at atmospheric pressure and 100°C, then the volume of gas required per liter of seawater can be calculated from the Ideal Gas Law:

$$V = \frac{nRT}{P} = \frac{(3.87 \cdot 10^{-3})(82.05 \cdot 10^{-3})(373)}{0.2} = 0.59 \text{ L}. \qquad \blacksquare$$

EXAMPLE 12*   A geothermal energy system pumps water at about 200°C from a deep well in the ground. Analysis of a sample cooled under 21 kg/cm² (approximately 21 atm) pressure gave (in mole/kg):

$$[Ca^{++}] = 5 \cdot 10^{-5}, \qquad\qquad [SO_4^{=}] = 6.2 \cdot 10^{-4},$$
$$[Mg^{++}] = 1 \cdot 10^{-5}, \qquad\qquad [HCO_3^{-}] = 3.72 \cdot 10^{-3},$$
$$[Na^{+}] = 4.67 \cdot 10^{-3}, \qquad\qquad [Cl^{-}] = 1.9 \cdot 10^{-4}.$$
$$[K^{+}] = 3.5 \cdot 10^{-4},$$

The positive and negative charges balance well enough, so that no estimation of missing ions seems necessary. From this list, $I = 0.0058$, which would give $f(I) = 0.07$ at 25°C, but at 200°C, $f(I) = 0.051$. Although the activity coefficients get closer to unity at higher temperatures, the ion pairs grow stronger and more than compensate (see Problem 26). Because of the high CO_2 content of the water, the pumps, well pipes, and heat exchangers must be kept under 10 to 20 atm pressure. The well operators would like to use the lowest pressure possible, because

---

* This example is based on notes by H. D. Holland.

that means longer equipment life and less chance for serious accidents if failure occurs. However, in tests run at lower pressure, calcium carbonate scale formed in the pipes. Use the data available to you to estimate the optimum operating pressure that will minimize both scale formation and corrosion.

If the water is just saturated with $CaCO_3$, the Langelier index is zero. Equations (6.32) and (6.33) with $L = 0$ give

$$pH = pK_{a2}^0 - pK_{s2}^0 + 2.5f(I) - \log A - \log[Ca^{++}]. \tag{6.38}$$

The partial pressure of CO_2 is determined from (4.4) or the equivalent:

$$[H^+][HCO_3^-] = K_{a1}K_H P,$$

or, with $[HCO_3^-] = A$,

$$\log P = pK_{a1}^0 + pK_H^0 - pH + \log A - 0.5f(I). \tag{6.39}$$

Combining (6.39) with (6.38) to eliminate pH, we get

$$\log P = -pK_{a2}^0 + pK_{s0}^0 + pK_{a1}^0 + pK_H^0 - 3.0f(I) + 2\log A + \log[Ca^{++}]. \tag{6.40}$$

Next come the numerical values: $A = [HCO_3^-] = 3.72 \cdot 10^{-3}$, $[Ca^{++}] = 5 \cdot 10^{-5}$, and $f(I) = 0.051$. From the data (Table 2.2) at 200°C, $pK_H^0 = 2.05$, $pK_{a1}^0 = 7.08$, $pK_{a2}^0 = 10.71$, $pK_{s0}^0 = 11.62$. Substituting in (6.40) gives $P = 5.34$ atm. At 200°C, you can find the vapor pressure of water in the standard steam tables (225.5 psia or 210.8 psig* at 392°F) to be 15.34 atm. Thus the total pressure of $H_2O + CO_2$ will be 20.68 atm.

Of course, a number of factors make this calculation uncertain. Most important are the activity coefficients and ion-pair effects, which tend to reduce the effective $[Ca^{++}]$ and hence reduce the equilibrium value of P. (See Eq. (6.40).) These same problems affect the measurement of equilibrium constants, and so these constants are less accurate than the 25°C data. Finally, the optimum degree of saturation may not be $L = 0$, and worse still, the scale may be partly $CaSO_4$, which is not taken into account here. ∎

THE LIME–SODA WATER SOFTENING PROCESS

Calcium and magnesium ion, which cause hardness in water, are normally accompanied by some carbonate alkalinity. If pH is raised, part of the calcium will precipitate as $CaCO_3$ and part of the magnesium as $Mg(OH)_2$, thereby softening the water. To accomplish this cheaply on a large scale, pH is raised with lime (CaO or $Ca(OH)_2$, as in Example 6) and the excess calcium introduced is precipitated as $CaCO_3$ by the addition of Na_2CO_3. In the course of this process, the pH

---

* The abbreviation psia stands for "pounds per square inch absolute" and psig stands for "pounds per square inch gauge." Gauge pressure (the practical measurement) is one atmosphere (14.696 lb/in^2) less than absolute pressure (the quantity used for gas law calculations).

of the water is raised and most of the magnesium, as well as calcium, is precipitated. After the precipitates have been flocculated, the clear supernatant is recarbonated with CO_2 (as in Example 3) to lower the pH to an acceptable level.

A simple stoichiometric approach predicts that optimum precipitation of both Ca^{++} and $CO_3^=$ as $CaCO_3$ occurs when the overall amount of Ca equals the overall amount of carbonate. If Q_1 is the number of moles of $Ca(OH)_2$ added and Q_2 is the number of moles of Na_2CO_3 added per liter, then

$$Q_1 + [Ca^{++}]^0 = Q_2 + C_T^0, \tag{6.41}$$

where $[Ca^{++}]^0$ is the number of moles of Ca per liter and C_T^0 the total carbonate concentration in the initial water. Equation (6.41) gives the difference between Q_1 and Q_2, but not either of them separately. Indeed, you might choose $Q_1 = 0$ if calcium were in excess and $Q_2 = 0$ if carbonate were in excess.

However, this approach would not take account of two important factors. First, Na_2CO_3 is substantially more expensive than $Ca(OH)_2$, and a good engineer would want to use as little as possible of the former. Second, the pH must be high enough to flocculate $CaCO_3$ (see p. 216) and to precipitate most of the Mg^{++}. This would not be the case if only enough $Ca(OH)_2$ were added to precipitate the existing Ca^{++}: pH would tend to be in the region around 8.3.

The solubility product of $Mg(OH)_2$ at 25°C is

$$K_{s0}^{0(Mg)} = [Mg^{++}][OH^-]^2 \gamma_{++}\gamma_-^2 = 10^{-10.7}, \tag{1.45}$$

or with the activity coefficients evaluated by the Davies equation:

$$K_{s0}^{Mg} = [Mg^{++}][OH^-]^2 = 10^{-10.7 + 3.0f(I)}. \tag{6.42}$$

As usual, $f(I)$ is given by the Davies equation (2.36). Thus, to keep $[Mg^{++}] < 10^{-4}$ at $I = 10^{-3}$, $\log[OH^-]$ must be greater than $\frac{1}{2}(-10.7 + 4.0 + 0.03) = -3.3$, and pH would be above 10.7. Raising the pH this high with $Ca(OH)_2$ alone would increase $[Ca^{++}]$, hence increase hardness and defeat part of the purpose; but an equimolar combination of $Ca(OH)_2$ and Na_2CO_3 can raise the pH without increasing $[Ca^{++}]$. A little additional Na_2CO_3, of course, will not do any harm, since it will suppress $[Ca^{++}]$ further, as you can see from the solubility product of $CaCO_3$: $[Ca^{++}] = K_{s0}/[CO_3^=]$.

STOICHIOMETRIC CALCULATIONS

In the traditional engineering approach* to this process, the carbonate species of the initial water are treated as separate reactants. Knowing two of the three quantities (alkalinity, total carbonate, pH), you can use Eq. (2.28) to obtain the other. For the present discussion, assume that pH is low enough (5 to 7), so that

---

* Fair, G. M., Geyer, J. C., and Okun, D. A. 1968. *Water and Wastewater Treatment*, vol. 2. New York: Wiley, pp. 29-10 to 29-14.

$[CO_3^=]$ is small compared to $[HCO_3^-]$ and

$$[HCO_3^-] = A + \cdots, \qquad [CO_2] = C_T - A + \cdots.$$

Step 1.   Add enough $Ca(OH)_2$ to react with CO_2*:

$$CO_2 + Ca(OH)_2 \rightarrow CaCO_3 + H_2O.$$

This will bring pH to about 8.

Step 2.   React the preexisting Ca^{++} and HCO_3^- with $Ca(OH)_2$:

$$Ca^{++} + 2HCO_3^- + Ca(OH)_2 \rightarrow 2CaCO_3 + H_2O.$$

Depending on whether there is excess Ca^{++} or excess HCO_3^- at this step, the calculations will be different. Consider first the case where HCO_3^- is in excess.

Step 3.   If there is no Mg^{++} present, the excess HCO_3^- can be stoichiometrically reacted with $Ca(OH)_2$ (if Mg^{++} is present, go to Step 5):

$$HCO_3^- + Ca(OH)_2 \rightarrow CaCO_3 + H_2O + OH^-.$$

Step 4.   If there is excess Ca^{++} in Step 2, then this must be reacted with Na_2CO_3:

$$Ca^{++} + Na_2CO_3 \rightarrow CaCO_3 + 2Na^+.$$

(If Mg^{++} is also present, go to Step 6.)

Step 5.   If there is excess HCO_3^- and also Mg^{++}, then they can be removed together by addition of $Ca(OH)_2$:

$$Mg^{++} + 2HCO_3^- + 2Ca(OH)_2 \rightarrow 2CaCO_3 + Mg(OH)_2 + 2H_2O.$$

If all the Mg^{++} is removed by this reaction but there is still HCO_3^- in excess, reaction with $Ca(OH)_2$ (Step 3) will take care of the excess. Note that excess HCO_3^- is not compatible with full precipitation of $CaCO_3$ and $Mg(OH)_2$.

Step 6.   If Mg^{++} is in excess, it can be removed by addition of equimolar $Ca(OH)_2$ and Na_2CO_3:

$$Mg^{++} + Ca(OH)_2 + Na_2CO_3 \rightarrow CaCO_3 + Mg(OH)_2 + 2Na^+.$$

This series of decisions can be seen more clearly on a logical flow diagram (Fig. 6.5), similar to a flowchart used in computer programming. Indeed, a simple computer program could be written (see Problem 20) to find the total amount of lime Q_1 and the total amount of sodium carbonate Q_2 as the sum of the partial quantities Q_{1k} and Q_{2k} given in the diagram, where k is the index giving the step. For example, Q_{13} is the amount of $Ca(OH)_2$ needed for Step 3, Q_{24} is the amount of Na_2CO_3 needed for Step 4, etc.

---

* Of course, it is impossible to convert all CO_2 to $CaCO_3$ because at equilibrium there will be some finite $[CO_2]$ at all pH values. However, you can see from Fig. 4.1 or 2.4 that, at pH = 8, $[CO_2]$ is about 2% of carbonate alkalinity.

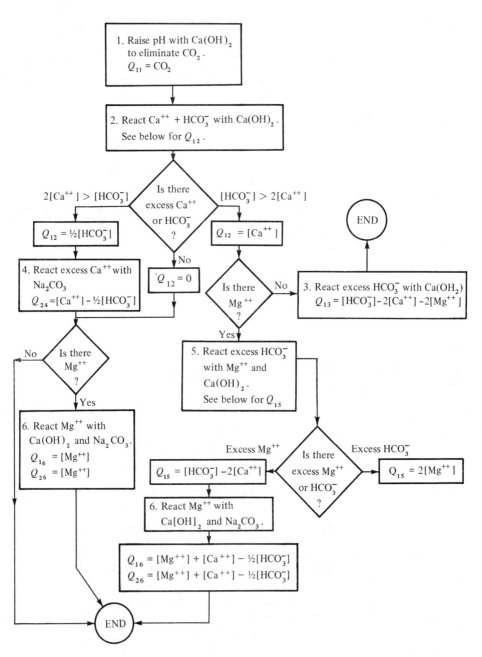

Fig. 6.5. Logical decision structure of the stoichiometric model for the lime–soda water softening process. This diagram is used in the text to obtain the algebraic results given by Eqs. (6.43) to (6.48).

Looking at the flow diagram, you can see that there are three main paths for which Mg^{++} is not zero, and the totals Q_1 and Q_2 are given for each of these by the sum of appropriate terms in the diagram:

Path A.   Both Ca^{++} and Mg^{++} in excess over HCO_3^-:

$$2[Ca^{++}]^0 + 2[Mg^{++}]^0 > [HCO_3^-]^0.$$

$$Q_1 = Q_{11} + Q_{12} + Q_{16} = [CO_2]^0 + [Mg^{++}]^0 + \frac{1}{2}[HCO_3^-]^0, \qquad (6.43)$$

$$Q_2 = Q_{24} + Q_{26} = [Ca^{++}]^0 + [Mg^{++}]^0 - \frac{1}{2}[HCO_3^-]^0. \qquad (6.44)$$

Path B. HCO_3^- in excess over both Ca^{++} and Mg^{++}:

$$[HCO_3^-]^0 > 2[Ca^{++}]^0 + 2[Mg^{++}]^0.$$

$$Q_1 = Q_{11} + Q_{12} + Q_{15} + Q_{13} = [CO_2]^0 + [HCO_3^-]^0 - [Ca^{++}]^0$$
$$= C_T - [Ca^{++}] + \cdots, \qquad (6.45)$$

$$Q_2 = 0. \qquad (6.46)$$

Path C. HCO_3^- in excess over Ca^{++}: $[HCO_3^-]^0 > 2[Ca^{++}]^0$, but not over Mg^{++} and Ca^{++} in total.

$$Q_1 = Q_{11} + Q_{12} + Q_{15} + Q_{16} = [CO_2]^0 + [Mg^{++}]^0 + \frac{1}{2}[HCO_3^-]^0, \qquad (6.47)$$

$$Q_2 = Q_{26} = [Ca^{++}]^0 + [Mg^{++}]^0 - \frac{1}{2}[HCO_3^-]^0. \qquad (6.48)$$

Note two important simplifications. First, both Path A and Path C give the same values for Q_1 and Q_2. Second, if HCO_3^- is in excess over both Ca^{++} and Mg^{++} (Path B), no Na_2CO_3 is required, and the amount of $Ca(OH)_2$ required is independent of $[Mg^{++}]$.

The quantity $2[Ca^{++}] + 2[Mg^{++}] - A = 2Q_2 + \cdots$ is often referred to as *noncarbonate hardness. Carbonate hardness* can be partially removed by boiling, which releases CO_2 and raises pH, permitting $CaCO_3$ to precipitate:

$$Ca^{++} + 2HCO_3^- \rightarrow CaCO_3 \text{ (s)} + CO_2 \text{ (g)}.$$

FORMULATION AS AN EQUILIBRIUM PROBLEM

It is conventional for engineers to add a little excess of both lime and soda, to speed up precipitation, and to aid in coagulation of the precipitate (see p. 216). There is no doubt that kinetic factors are important in running the process efficiently, but it is also useful to know what the minimum amount of residual hardness might be after the process has come to equilibrium.

In the following, Q_1 is the amount of $Ca(OH)_2$ added, Q_2 the amount of Na_2CO_3 added, P_1 the amount of $CaCO_3$ precipitated, and P_2 the amount of $Mg(OH)_2$ precipitated. Superscript zeros on concentration variables indicate the values for the initial water; those without superscripts indicate the concentrations at equilibrium.

Mass balances on Ca, Mg, Na, and carbonate yield

$$[Ca^{++}]^0 + Q_1 = [Ca^{++}] + P_1, \tag{6.49}$$

$$[Mg^{++}]^0 = [Mg^{++}] + P_2, \tag{6.50}$$

$$[Na^+]^0 + 2Q_2 = [Na^+], \tag{6.51}$$

$$C_T^0 + Q_2 = C_T + P_1. \tag{6.52}$$

The charge balance is

$$2[Ca^{++}] + 2[Mg^{++}] + [Na^+] + [H^+] = [HCO_3^-] + 2[CO_3^=] + [OH^-]. \tag{6.53}$$

An equation of the same form as (6.53), but with all terms superscripted zero, also holds. These six equations can be combined to eliminate five independent variables. Since P_1 and P_2 are not easily measured, set aside Eq. (6.50) (for evaluating P_2 later, if desired), and subtract (6.52) from (6.49) to get rid of P_1:

$$[Ca^{++}]^0 + Q_1 - C_T^0 - Q_2 = [Ca^{++}] - C_T. \tag{6.54}$$

$[Na^+]^0$ and $[Na^+]$ can be obtained from (6.53) and substituted in (6.51) to yield:

$$2Q_2 + [HCO_3^-]^0 + 2[CO_3^=]^0 + [OH^-]^0 - 2[Ca^{++}]^0 - 2[Mg^{++}]^0 - [H^+]^0$$
$$= [HCO_3^-] + 2[CO_3^=] + [OH^-] - 2[Ca^{++}] - 2[Mg^{++}] - [H^+]. \tag{6.55}$$

It would seem logical now to eliminate C_T, which is not easily measured directly, but this leads to a rather cumbersome set of equations. Instead, eliminate Q_2 by substituting (6.55) in (6.54):

$$2[CO_2]^0 + [HCO_3^-]^0 + 2[Mg^{++}]^0 + [H^+]^0 - [OH^-]^0 - 2Q_1$$
$$= 2[CO_2] + [HCO_3^-] + 2[Mg^{++}] + [H^+] - [OH^-]. \tag{6.56}$$

The equilibria for carbonate ((2.37), (2.39), (2.42)) and the solubility product of $Mg(OH)_2$ (Eq. (6.42)) can be substituted in (6.56) to give:

$$C_T^0 \frac{2(10^{-2pH^0}) + K_1'(10^{-pH^0})}{y^0} + 2[Mg^{++}]^0 - K_w'(10^{+pH^0}) + 10^{-pH^0} - 2Q_1$$

$$= C_T \frac{2(10^{-2pH}) + K_1'(10^{-pH})}{y} + \frac{2K_{s0}^M(10^{-2pH})}{K_w'^2} - K_w'(10^{+pH}) + 10^{-pH}, \tag{6.57}$$

where $y = K_1'K_2' + K_1'10^{-pH} + 10^{-2pH}$. The left-hand side of (6.56) or (6.57) contains only variables relating to the initial conditions (C_T^0, pH^0, $[Mg^{++}]^0$, Q_1), while the right-hand side contains only variables relating to the equilibrium conditions.

C_T can be obtained in terms of the initial conditions and equilibrium pH by using the solubility product of $CaCO_3$ (Eq. (4.1)) together with (6.54):

$$\frac{K_{s0}y}{K_1'K_2'} = C_T([Ca^{++}]^0 + Q_1 - Q_2 + C_T - C_T^0). \tag{6.58}$$

This is a quadratic in C_T that can be solved explicitly by using the quadratic formula:

$$C_T = \frac{1}{2}(Q_2 - Q_1 + C_T^0 - [Ca^{++}]^0)$$

$$+ \frac{1}{2}\left[(Q_2 - Q_1 + C_T^0 - [Ca^{++}]^0)^2 + \frac{4K_{s0}'y}{K_1'K_2'}\right]^{1/2}. \tag{6.59}$$

The positive square root was chosen, since the first group of terms in parentheses are all independent variables and their sum could be positive, negative, or zero. If this group were zero, $C_T = (K_{s0}y/K_1'K_2')^{1/2}$, which must be positive.

Equations (6.57) and (6.59) together give the final equilibrium pH implicitly as a function of the independent initial parameters (C_T^0, $[Ca^{++}]^0$, $[Mg^{++}]^0$, pH^0) and the quantities of reagents added (Q_1 and Q_2).

Now examine what equilibrium composition is reached when the amounts of reagent added are given by the stoichiometric model (Path A or C) of the previous section. These equations for Q_1 and Q_2 result in great simplification. Subtract (6.43) from (6.44) and note that

$$C_T^0 = [CO_2]^0 + [HCO_3^-]^0 + [CO_3^=]^0 \tag{2.18}$$

to get

$$Q_2 - Q_1 + C_T^0 - [Ca^{++}]^0 = [CO_3^=]^0 = C_T^0\frac{K_1'K_2'}{y^0}. \tag{6.60}$$

This result transforms (6.59) to the simpler form

$$C_T = \frac{1}{2}\frac{C_T^0 K_1'K_2'}{y^0} + \frac{1}{2}\left[\left(\frac{C_T^0 K_1'K_2'}{y^0}\right)^2 + \frac{4K_{s0}y}{K_1'K_2'}\right]^{1/2}. \tag{6.61}$$

For the normal range of variables, with initial pH near 7 and final pH near 11, Eq. (6.61) can be approximated by the amazingly simple

$$C_T = K_{s0}^{1/2} + \cdots. \tag{6.62}$$

Equations (6.56) and (6.57) also become much simpler. Substituting for Q_1 from (6.43) in (6.56) gives

$$[H^+]^0 - [OH^-]^0 = [HCO_3^-]^0 + 2[CO_2] + 2[Mg^{++}]$$
$$+ [H^+] - [OH^-], \tag{6.63}$$

in which all the initial parameters have canceled out! If the initial pH is near 7 and the final pH near 11, the terms $[H^+]^0$, $[OH^-]^0$, $[H^+]$, and $[CO_2]$ can all be neglected compared to the others, and the equilibria can be substituted, resulting in

$$K'_w(10^{+\text{pH}}) = K_{s0}^{1/2} \frac{10^{-\text{pH}}}{K'_2} + 2K_{s0}^M \frac{10^{-2\text{pH}}}{K_w'^2} + \cdots. \tag{6.64}$$

This equation is still cubic in $[\text{H}^+]$ but is easily solved by trial and error in the form

$$\text{pH} = \frac{1}{2}\log\left(\frac{K_{s0}^{1/2}}{K'_2 K'_w} + 2K_{s0}^M \frac{10^{-\text{pH}}}{K_w'^3}\right) + \cdots. \tag{6.65}$$

EXAMPLE 13   Let $C_T^0 = 10^{-3.0}$, $\text{pH}^0 = 7.5$. Equation (6.62) with $pK_{s0} = 8.3$ yields $C_T = 10^{-4.15}$, with the omitted terms in (6.61) being of the order of 10^{-6}. Equation (6.66) (with $pK'_2 = 10.3$, $pK_w = 14.0$, and $pK_{s0}^M = 10.7$) yields

$$\text{pH} = \frac{1}{2}\log(10^{+20.15} + 10^{+31.6-\text{pH}}) + \cdots.$$

Substituting $\text{pH} = 10.5$ gives a right-hand side value of 10.57; substituting $\text{pH} = 10.57$ gives a right-hand side value of 10.54. A third iteration gives $\text{pH} = 10.56$, which satisfies the equation. The main terms in (6.64) are about $10^{-3.5}$; the neglected terms that occur in (6.63) are about 10^{-6}. (Increasing ionic strength decreases pK_{s0}, pK'_2, pK_w, and pK_{s0}^M, and increases the pH calculated from Eq. (6.65); see Problem 23.) Having obtained the final pH, you can now calculate the residual hardness, which is the sum of calcium and magnesium concentrations:

$$[\text{Ca}^{++}] + [\text{Mg}^{++}] = \frac{K_{s0}}{[\text{CO}_3^=]} + \frac{K_{s0}^M}{[\text{OH}^-]^2}. \tag{6.66}$$

With $C_T = K_{s0}^{1/2}$,

$$[\text{Ca}^{++}] + [\text{Mg}^{++}] = \frac{K_{s0}^{1/2} y}{K'_1 K'_2} + \frac{K_{s0}^M(10^{-2\text{pH}})}{K_w'^2}$$

$$= 10^{-3.96} + 10^{-3.80} = 10^{-3.57}. \tag{6.67}$$

This corresponds to a residual hardness of 26 mg/L (as $CaCO_3$)—a typical goal for water softening. Note that this result is *independent* of the initial hardness, provided the dosage of lime and soda is calculated by the stoichiometric model (Eqs. (6.43) and (6.44)). If the initial $[\text{Mg}^{++}]$ is too small to precipitate, then (6.67) overestimates the Mg contribution to hardness.   ∎

At the beginning of Chapter 4 you saw that ion pairs such as $[\text{CaCO}_3^0]$ and $[\text{CaHCO}_3^+]$ form a small, but sometimes significant, portion of the total dissolved calcium, and hence of the hardness. Depending on which of the various published thermodynamic data you choose, however, you will get different estimates of the minimum hardness achievable in the presence of solid $CaCO_3$. Figures 4.1 and 4.2 give one set of estimates. At $P_{CO_2} = 10^{-3.5}$ atm, $[CaCO_3^0] > [Ca^{++}]$ for pH > 9.3, and above that $C_{Ca} = [CaCO_3^0] = 10^{-5.4}$ (Eq. (4.8)). At $P_{CO_2} = 1.0$ atm, $[CaHCO_3^+]$ exceeds $[Ca^{++}]$ for pH > 6.6, and $[CaCO_3^0] > [CaHCO_3^+]$ for pH > 8.5. Above that pH, $C_{Ca} = 10^{-5.4}$.

This high-pH limit, which is independent of pH and P_{CO_2}, corresponds to only 0.4 mg/L hardness as $CaCO_3$, which is certainly an acceptable value. However, the formation of ion pairs with any of the other ions present in the water, especially sulfate, can increase the solubility of $CaCO_3$. In addition, even careful laboratory studies are not all in good agreement. At the other extreme, one group* recently calculated the limiting hardness from a different set of data ($K_1 = 10^{+4.45}$ for $CaCO_3$ ion-pair formation) than I used above ($K_1 = 10^{+3.1}$), and obtained $C_{Ca} = 13.5$ mg/L as $CaCO_3$.

RECARBONATION

The softened water may be treated with CO_2 to lower the pH after flocculation. Recall from Chapter 2 and examples on pp. 190–192 that addition or withdrawal of CO_2 does not change the alkalinity; thus the final product has the same alkalinity as the equilibrium water:

$$A = [Na^+] + 2[Mg^{++}] + 2[Ca^{++}].$$

In terms of the final pH and initial parameters, provided the dosage Q_2 of Na_2CO_3 is calculated from (6.44), this becomes

$$A = [Na^+]^0 + 2[Ca^{++}]^0 + 2[Mg^{++}]^0 - C_T^0 \frac{K_1'(10^{-pH^0})}{y^0}$$

$$+ \frac{2K_{s0}^{1/2}y}{K_1'K_2'} + \frac{2K_{s0}^M(10^{-2pH})}{K_w'^2}, \tag{6.68}$$

where the last two terms are as given by (6.67).

Here C_T is increased by the amount of CO_2 added. If all $CaCO_3$ has been removed before recarbonation, this will be simply

$$C_T^{final} = C_T + \Delta[CO_2] = K_{s0}^{1/2} + \Delta[CO_2]. \tag{6.69}$$

Knowing A and C_T, we can calculate the final pH by using equations (2.29) and (3.11).

$$A = C_T \frac{K_1'(10^{-pH}) + 2K_1'K_2'}{y} + K_w'(10^{+pH}) + 10^{-pH}.$$

---

* Cadena, F., Midkiff, W. S., and O'Connor, G. A. 1974. *J. Am. Water Works Assoc.* 66:524–526. They used data obtained by Martynova, O. J., Vasina, L. G., Pozdnyakova, S. A. 1971. *Dokl. Akad. Nauk SSSR* 202:6.

More commonly, a target pH will be chosen and C_T^{final} calculated; then Eq. (6.69) gives the $\Delta[CO_2]$ needed.

CHEMICAL COAGULATION AND FLOCCULATION

Removal of suspended material in a raw water before it is processed for a municipal water supply, clarification of the effluent from a secondary sewage treatment plant, phosphorus removal by alumina or lime in tertiary treatment of effluent— all these require precise adjustment of pH so as to achieve the optimum degree of coagulation. Discussion of colloid chemistry and the details of the coagulation process are beyond the scope of this book, but interested readers are referred to the literature.*

EXAMPLE 14   Precipitation of phosphorus from wastewater as a calcium phosphate layer on the surface of calcium carbonate crystals can be modeled as an equilibrium between calcite and hydroxyapatite (Stumm, W. and Morgan, J. J. 1981. *Aquatic Chemistry*. New York: Wiley, p. 284):

$$10CaCO_3 \text{ (s)} + 2H^+ + 6HPO_4^= + 2H_2O \rightleftharpoons Ca_{10}(PO_4)_6(OH)_2 \text{ (s)} + 10HCO_3^-,$$

with equilibrium constant approximately

$$K_{\text{phos}} = 10^{+32} = \frac{[HCO_3^-]^{10}}{[H^+]^2[HPO_4^=]^6}. \tag{6.70}$$

Approximating $[HCO_3^-]$ by alkalinity A, and $[HPO_4^=]$ by total phosphorus C_P, the equilibrium phosphorus level predicted by the model depends on both pH and alkalinity; we get

$$\log C_P = +\frac{5}{3}\log A + \frac{1}{3}pH - 5.33. \tag{6.71}$$

For alkalinity 3×10^{-3} and pH $= 9.8$ (see below for why these values were chosen), Eq. (6.71) gives

$$\log C_P = -4.20 + 3.27 - 5.33 = -6.27.$$

Since the atomic weight of phosphorus is 30.97 g/mole, this corresponds to $C_P = 1.7 \cdot 10^{-2}$ mg/L.

Lower alkalinity and lower pH will tend to decrease the residual dissolved phosphorus. A more complicated set of equations would be required to take account of the acid–base equilibrium of HCO_3^- and $HPO_4^=$.   ■

---

* Fair, G. M., Geyer, J. C., and Okun, D. A. 1968. *Water and Wastewater Treatment*, vol. 2. New York: Wiley; Weber, W. 1972. *Physicochemical Processes for Water Quality Control*. New York: Wiley.

However, particulate phosphorus will remain in suspension at concentrations of several milligrams per liter for hours or even days at pH 9 to 10. It is thus not the precipitation of dissolved phosphorus that is critical, but the flocculation of the precipitate. This does not occur rapidly until pH is about 11.5, and requires a substantial excess of lime. According to the model used in Example 14, C_P would increase by about a factor of ten for such a pH change.

Fortunately, the addition of a little magnesium (from seawater, for example) greatly increases the rate of flocculation and hence permits the use of a lower pH. This effect is shown in Fig. 6.6.* Addition of 15% seawater (200 ppm Mg^{++}) together with 100 ppm $Ca(OH)_2$ (alkalinity increase $2.7 \cdot 10^{-3}$ M) leaves residual phosphorus in the supernatant of 1.5 mg/L at pH = 9.8 (still 100 times the value we calculated above by the equilibrium model). But in contrast, addition of the same amount of $Ca(OH)_2$ without magnesium leaves a residual phosphorus concentration of 4.6 mg/L under the same settling times and stirring conditions. The mechanism of this effect is not yet understood, but to reduce phosphorus to 1.5 mg/L with $Ca(OH)_2$ alone would have required about 300 mg/L of $Ca(OH)_2$ and would have raised the pH to 11.4.

One important lesson to be learned from this example is that equilibrium is rarely reached with respect to precipitation reactions in practical systems, and questions of the rate of coagulation, settling, or crystallization become much more important than the thermodynamic limiting concentration. However, acid–base reactions (such as protonation of the carbonate and phosphate species) occur as rapidly as solutions can be mixed, and so equilibrium calculations of pH for homogeneous solutions (when proper account is taken of the ionic strength and all the chemical species) can be quite accurate.

REMOVAL OF CO₂ FROM PROCESS GAS

As a final example, consider the removal of CO_2 from a gas stream by passing it through an alkaline solution. In the early days of gas analysis (before gas chromatography), concentrated KOH or NaOH was used to remove CO_2 quantitatively. However, NaOH supported on inert material as Ascarite® is an expensive reagent and eventually becomes Na_2CO_3 and must be discarded. A concentrated solution of Na_2CO_3 or K_2CO_3 is less expensive and can be regenerated because the reaction

$$Na_2CO_3 + CO_2 + H_2O \rightleftharpoons 2NaHCO_3$$

is reversible within normally accessible temperature and pressure ranges. At lower temperatures and higher CO_2 partial pressures, the reaction proceeds to the right; at higher temperatures and lower pressure, it proceeds to the left.

---

* Vrale, L. 1978. *Prog. Water Technol.* 10:645–656.

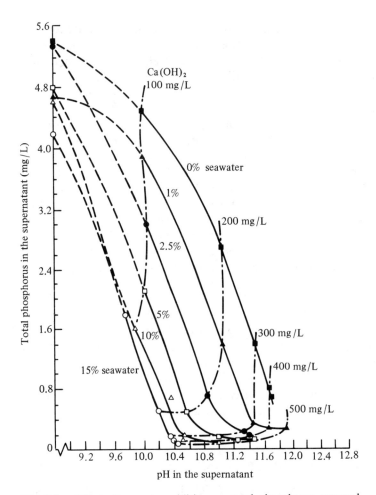

Fig. 6.6.   Effect of seawater addition on total phosphorus removal by lime from wastewater. Contours of constant $Ca(OH)_2$ addition and percent seawater addition show that addition of 15% seawater and 100 mg/L lime can reduce the residual phosphorus to the same level (1.6 mg/L) that would be achieved with 300 mg/L $Ca(OH)_2$ in fresh water. The optimum phosphorus removal is achieved with about 5% seawater and 300 mg/L lime, resulting in about 0.2 mg/L residual phosphorus. (From Vrale, L., op. cit.)

One application that is of current interest is the use of low-grade fossil fuels to produce hydrogen:

$$2H_2O + CH_n \rightarrow \left(2 + \frac{n}{2}\right)H_2 + CO_2.$$

This "reformer gas" can be used to produce electricity in a fuel cell if the CO_2 is first removed, since the fuel cell electrolyte is alkaline, and any substantial quantity of CO_2 in the hydrogen input would greatly diminish the performance of the fuel cell. One energy-efficient process for CO_2 removal is the *hot-carbonate process.**

EXAMPLE 15   A gas stream from the reformer containing 30% CO_2, 70% H_2, and some traces of water and methane, at 300 psig (21.4 atm absolute), contacts a 35%-by-weight K_2CO_3 solution at 230°F (110°C). (The Bureau of Mines reported that 3 to 5 SCF (standard cubic feet) of CO_2 were absorbed per gallon of solution: 1.0 to 1.7 mole/L; in plant operation, 2.5 SCF of CO_2 per gallon at 300 psig was more typical.) The spent solution, typically 60–70% converted to bicarbonate, is reduced to atmospheric pressure (2–3 psig) in a flash druin at the top temperature of the regenerator (210°F), and the remainder of the CO_2 is removed by steam stripping. The regenerated solution is about 30% bicarbonate. Using constants available in Table 2.2, estimate the equilibrium amount of CO_2 that could be expected in the gas exiting from the absorber when it is 10% converted to bicarbonate; and the minimum amount of bicarbonate that might be expected on regeneration without steam stripping.

The equilibria and mass balances are familiar; at high pH,

$$C_T = [HCO_3^-] + [CO_3^=] + \cdots = K_H P\left(\frac{K_{a1}}{[H^+]} + \frac{K_{a1}K_{a2}}{[H^+]^2}\right) + \cdots, \quad \text{(2.18)}$$

$$A = [K^+] = [HCO_3^-] + 2[CO_3^=] + [OH^-] + \cdots$$

$$= K_H P\left(\frac{K_{a1}}{[H^+]} + \frac{2K_{a1}K_{a2}}{[H^+]^2}\right) + \frac{K_w}{[H^+]} + \cdots. \quad \text{(2.16)}$$

Recall that C_T increases with P, but A remains constant as CO_2 is added or removed; $[H^+]$ and pH depend on the ratio of $[HCO_3^-]$ to $[CO_3^=]$, which is given above as "10% converted":

$$[HCO_3^-] = 0.1 C_T.$$

Making use of (2.18) and (2.3), you get

$$[H^+] = 0.11 K_{a2}. \quad \text{(6.72)}$$

Substitute (6.72) in (2.16) and solve for P:

$$P = \left(A - 9.0\frac{K_w}{K_{a2}}\right)\frac{K_{a2}}{171 K_{a1} K_H}. \quad \text{(6.73)}$$

---

* Giner, J. and Swette, L. 1975. *Evaluation of the Feasibility of Low-cost Carbon Dioxide Removal/Transfer Methods for Fuel Cell Application.* Palo Alto, Calif.: Electric Power Research Institute, Report No. EPRI 391.

The easiest part of the numerical work is looking up the constants at $100°C$ in Table 2.2: $pK_H^0 = 1.99$, $pK_{a1}^0 = 6.45$, $pK_{a2}^0 = 10.16$, and $pK_w^0 = 12.3$. The concentration of K_2CO_3 gives both A and the ionic strength; 35% by weight means $35/138.2$ moles in 65 cm^3 water or 3.90 mole/L. Therefore $[K^+] = A = 7.80$ mole/L, and

$$I = \frac{1}{2}([K^+] + [HCO_3^-] + 4[CO_3^=])$$

$$= \frac{1}{2}(7.80 + 0.39 + 14.04) = 11.1 \text{ mole/L}.$$

It is clear that at such high ionic strengths, the Davies equation will not give any more accurate estimate of activity coefficients than neglecting them entirely will. What else could you do? I looked at some estimates by Garrels and Christ* (Fig. 6.7) of the activity coefficients of carbonate and bicarbonate ions at room temperature. With a little imagination, you might decide that $\gamma_- = 0.6$ and $\gamma_= = 0.1$ at $I = 11$ are not ridiculous estimates. In any case, I will use them for the following calculations. When the activity coefficients are made explicit in (6.73), the numerical result can be obtained:

$$P = \left(A - 9.0 \frac{K_w^0}{K_{a2}^0} \frac{\gamma_=}{\gamma_-^2}\right) \frac{K_{a2}^0 \gamma_-^2}{171 K_{a1}^0 K_H^0 \gamma_=}$$

$$= (7.80 - 0.018)(10^{-3.40}) = 3.12 \cdot 10^{-3} \text{ atm.}$$

This represents better than 99.95% removal of CO_2 from the original partial pressure (30% of 21.4 atm is 6.42 atm) in the reformer gas.

The final composition of the electrolyte after using steam to purge CO_2 from the absorber will depend on the CO_2 content of the steam and the kinetics of the process. Without the steam purge, however, regeneration is not very effective. Set $P = 1$ atm, and obtain pH from (2.16):

$$A = K_H^0 P \left(K_{a1}^0 \frac{10^{+pH}}{\gamma_-} + 2K_{a1}^0 K_{a2}^0 \frac{10^{+2pH}}{\gamma_=}\right)$$

or

$$10^{+9.11} = 10^{+pH} + 10^{+2pH-9.08}$$

The two terms in pH are about equal, so the quadratic formula is needed to get pH = 8.89. Then (2.22) gives the relative amount of bicarbonate:

$$\frac{[HCO_3^-]}{C_T} = \frac{10^{-pH}}{10^{-pH} + K_{a2}\gamma_-/\gamma_=} = 10^{-0.12} = 0.756.$$

In other words, atmospheric pressure of CO_2 is in equilibrium with a solution containing 25% carbonate and 75% bicarbonate. Clearly, the purge is an important part of the process. ∎

* Garrels, R. M. and Christ, C. L. 1965. *Solutions, Minerals and Equilibria*. New York: Harper & Row, p. 104.

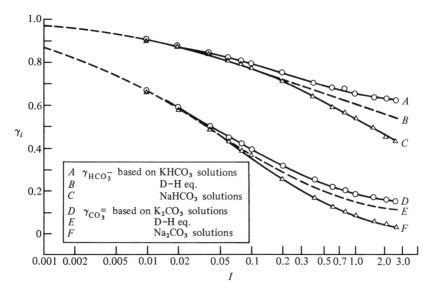

Fig. 6.7.   Activity coefficients estimated for bicarbonate and carbonate ions. Curves *A*, *C*, *D*, and *F* are based on experimental data from potassium and sodium salts (Walker, A. C. Bray, U. B. and Johnston, J. 1927. *J. Am. Chem. Soc.* 49:1255); curves *B* and *E* are based on the Debye–Hückel equation in the form

$$-\log \gamma_i = \frac{Az_i^2 I^{1/2}}{1 + a_i B I^{1/2}}$$

with $A = 0.5085$ and $B = 0.3281 \cdot 10^8$ cm^{-1} at 25°C. For HCO_3^-, $z_i = 1$ and $a_i = 4.0 \cdot 10^{-8}$ cm; for $CO_3^=$, $z_i = 2$ and $a_i = 4.5 \cdot 10^{-8}$ cm. (From Garrels, R. M. and Christ, C. L. 1965. *Solutions, Minerals, and Equilibria*. New York: Harper & Row, p. 104.)

PROBLEMS

1. A process water analysis is reported as follows (after ASTM standard SIC-28)

Fe_T*	5 mg/L	Alkalinity	1000 mg/L
Mn^{++}	2 mg/L		(as $CaCO_3$)
Ca^{++}	200 mg/L	Total dissolved	
Mg^{++}	100 mg/L	solids	2500 mg/L
$SO_4^=$	650 mg/L	pH	6.6
Cl^-	500 mg/L	Temperature	30°C

Estimate a value for ionic strength and the possible error in that value due to incomplete specification of the analysis.

---

* At pH = 6.6, 10^{-4} M oxidized Fe is present primarily as $Fe(OH)_3$ with a little $Fe(OH)_2^+$ (Stumm and Morgan, op. cit. p. 248). Reduced Fe is Fe^{++} and $FeOH^+$.

Answer: $R = 448$ (note alkalinity at pH $= 6.6$ is HCO_3^-; use 600 mg/L as HCO_3^-, not 1000 mg/L as $CaCO_3$) in Eq. (6.2) gives $I = 5.5 \cdot 10^{-2}$. If the residual charge balance $(1.94 \cdot 10^{-2}$ mole/L) is assumed to be Na^+, you get $I = 5.35 \cdot 10^{-2}$.

2. How much H_2SO_4 is required to reduce the alkalinity of the water in Problem 1 to 500 mg/L (as $CaCO_3$)? What is the final pH?

   Answer: $5.00 \cdot 10^{-3}$ mole of acid or 245 mg/L of H_2SO_4; pH $= 5.96$

3. How much NaOH is required to raise the pH of the water described in Problem 1 to pH $= 8.0$?

   Answer: $9.1 \cdot 10^{-3}$ mole/L

4. How would the answers to Problems 1, 2, and 3 be affected if only TDS, alkalinity, and pH were given? Estimate the errors resulting in the answers to Problems 2 and 3 if only TDS, alkalinity, and pH were known.

5. How much $Ca(OH)_2$ is required to raise the water described in Problem 1 to pH $= 8.0$? Does $CaCO_3$ precipitate?

   Answer: $4.56 \cdot 10^{-3}$ mole/L or 338 mg/L. Activity product is $10^{-6.61} \gg 10^{-8.57}$. $L = +1.96$. Yes.

6. How much CO_2 (as flue gas, 20% CO_2, 80% N_2) at 1 atm is required to reduce the pH of the water in Problem 1 to 5.0?

   Answer: Partial pressure of CO_2 must be greater than 5.9 atm to reach pH $= 5.0$.

7. What is the lowest pH value that can be reached by treating the water of Problem 1 with flue gas?

   Answer: With P $= 0.2$ atm, pH $= 6.47$.

8. What is the highest pH that can be reached by aerating the water of Problem 1 (with air containing 0.1% by volume CO_2)?

   Answer: pH $= 8.72$

9. What is the effect of treating the water of Problem 1 with limestone ($CaCO_3$)? Compare the results obtained by considering only calcium and carbonate species with those from an ion-pairing model. What effect does addition of limestone have on the hardness of this water?

   Answer: If you apply Langelier's equation to the initial solution, taking $A = [HCO_3^-]$ $= 10^{-2.00}$ and $[Ca^{++}] = 10^{-2.30}$ as given in Problem 1, you find $L = +0.14$ (supersaturated). Proceeding with the model which omits ion pairs, you find $D = 10^{-2.33}$, $D_2 = 10^{-1.73}$. These are invariant to addition of $CaCO_3$, and at saturation give pH $= 6.54$, $C_T = 10^{-1.85}$, $[Ca^{++}] = 10^{-2.35}$, and $L = 0$. If the strong ion pairs are included (at $I = 0.055$, $[MgSO_4] = 10^{+1.65}$ $[Mg^{++}][SO_4^=]$ and $[CaSO_4] = 10^{+1.60}$ $[Ca^{++}][SO_4^=]$) you find $[SO_4^=] = 10^{-2.32}$, $[Ca^{++}] = 10^{-2.38}$, $[Mg^{++}] = 10^{-2.47}$, and $L = +0.06$ for the initial conditions. When pH $= 6.57$, $L = 0$. Hardness is the sum of calcium and magnesium species, including the ion pairs. In the initial solution, hardness is 9.10×10^{-3} mole/L or 910 mg/L as $CaCO_3$, and at pH $= 6.57$, it is reduced to 880 mg/L. (See also Problem 14.)

10. Compare the cost of raising pH by use of $Ca(OH)_2$, Na_2CO_3, and NaOH, if the cost per ton is $75, $180, and $240, respectively. To what extent is this cost affected by the alkalinity,

hardness, etc. of the water to be treated? Use the water composition of an average U.S. river (Table 5.2) as the basis of your argument.

11. How does Langelier's index differ in principle from the geochemical *degree of saturation* defined by $[Ca^{++}][CO_3^=]/K_{so}$? What advantages does each have in practice?

12. Derive the rigorous form of Langelier's equation, analogous to Eq. (6.31), by using the ion-pairing model presented at the beginning of Chapter 4. Evaluate the difference between this equation and the Langelier formula (Eq. (6.32)) and plot as in Fig. 6.1. Compare the difference to that resulting from using $pK_{so}^0 = 8.34$ instead of 8.52.

13. Snoeyink and Jenkins (op. cit.) give the following equation for the definition of Langelier's index:

$$L.I. = pH - pK_{a2}^0 - pK_{so}^0 + p[Ca^{++}] + p[HCO_3^-] + \log \gamma_{Ca^{++}} + \log \gamma_{HCO_3^-},$$

where the equilibrium constants are *activity* or zero ionic strength constants. How does this differ from Eq. (6.32)? How much does it extend the pH range for which the concept is valid ($L.I. < 0$ means undersaturation and $L.I. > 0$ means supersaturation)? Sketch a curve of neglected terms as in Fig. 6.1.

14. For a water containing relatively large concentrations of divalent ions, the formation of ion pairs with Ca^{++} and $CO_3^=$ will increase the solubility of $CaCO_3$, and thus the prediction of the Langelier equation will be inaccurate. Using the ion-pairing constants from Table 5.7 (e.g., $[CaSO_4] = 10^{+2.31}[Ca^{++}][SO_4^=]$), calculate the maximum concentration of sulfate required to produce an error of 0.04 (i.e., $L = +0.04$ instead of zero for saturation). Perform a similar calculation for Mg^{++}.

15. Caldwell, D. H. and Lawrence W. B. (1953. *Ind. Eng. Chem.* 45:535–548) developed a graphical method, which was modified by Lowenthal and Marais (op. cit.), by using the variables

$$C_1 = [OH^-] - [H^+] - [HCO_3^-] - 2[CO_2],$$
$$C_2 = [HCO_3^-] + 2[CO_3^=] + [OH^-] - [H^+] - 2[Ca^{++}].$$

Using these as the x and y coordinates, curves of constant $[Ca^{++}]$, constant A, and constant pH were plotted as three grids. Derive the equation for these lines and plot one set for $pH = 8$, $A = 10^{-3}$, $[Ca^{++}] = 2 \cdot 10^{-3}$. Draw another pH line corresponding to saturation.

16. Show that the Caldwell–Lawrence diagram (see Problem 15) has the following properties:

Add or remove $CaCO_3$: no effect	($\odot$)
Add $Ca(OH)_2$: increase C_1; C_2 unchanged	($\uparrow$)
Add CO_2: decrease C_1; C_2 unchanged	($\downarrow$)
Add $CaCl_2$: C_1 unchanged; decrease C_2	($\leftarrow$)
Add Na_2CO_3: C_1 unchanged; increase C_2	($\rightarrow$)
Add $NaHCO_3$: decrease C_1 and increase C_2 equally	($\searrow$)
Precipitate $Mg(OH)_2$: decrease C_1 and C_2 equally	($\swarrow$)

Compare with the properties of the original Caldwell–Lawrence functions:

$$f = [OH^-] - [HCO_3^-] - [CO_2], \qquad \phi = [Ca^{++}] - A - [H^+].$$

17. What value of C_1 corresponds to the minimum calcium content?

    Answer: $C_1 = 0$

18. A water has pH $= 5.7$, alkalinity 80 mg/L (as $CaCO_3$), hardness 78 mg/L (as $CaCO_3$), $[Ca^{++}]$ 15 mg/L, and TDS $= 300$ mg/L. Calculate the Langelier index. Is it saturated with $CaCO_3$? How much CO_2 or $Ca(OH)_2$ is required to bring it to the point where it is barely saturated?

    Answer: At 25°C, $L = -2.83$ (unsaturated). Addition of $3.2 \cdot 10^{-3}$ mole/L $Ca(OH)_2$ (or 234 mg/L) will bring pH to 7.1, $[Ca^{++}]$ to $10^{-2.71}$, and L to zero.

19. What is the Langelier index for a water if 5 mg/L $CaCO_3$ would precipitate on nucleation? Assume the water remains supersaturated.

20. Develop a computer program in your favorite language (BASIC, PASCAL, etc.) to perform the operations diagrammed in Fig. 5.5. Compare with the answers to Eqs. (6.43)–(6.48).

21. Eliminate C_T between Eqs. (6.54) and (6.55) to obtain a single equation relating pH, Q_1, Q_2, and the initial conditions. Then substitute (6.43) and (6.44) to obtain an equation in the initial conditions and pH. Compare with (6.64).

22. A water with $C_T^0 = 10^{-3.0}$, pH $= 7.5$ was treated with lime and soda and the final pH was 10.55. This gave a residual hardness $[Ca^{++}] + [Mg^{++}] = 10^{-3.57}$ in Example 13 of this chapter (Eq. (6.67)) under the assumption of no appreciable ion pairs. Recalculate this example assuming total $SO_4^= = 10^{-3.0}$ and introducing $CaSO_4$ and $MgSO_4$ complexes (Table 5.7). Does $CaSO_4$ ($K_{s0}^0 = 10^{-4.6}$) precipitate?

23. Analyses of a water gave

$$[Ca^{++}] = 210 \text{ mg/L}, \qquad [Mg^{++}] = 35 \text{ mg/L},$$
$$\text{Alkalinity} = 850 \text{ mg/L as } CaCO_3, \qquad pH = 7.2.$$

Calculate the amount of lime and soda required to achieve the *stoichiometric softening condition*, and the residual hardness. How is the latter result affected if ion pairs are included in the calculations for saturated $CaCO_3$?

24. Sketch a diagram of $[Ca^{++}]$ and $[Mg^{++}]$ as a function of pH at $C_T = 10^{-2}$ M (including carbonate in solid phases). Possible solid phases and their solubility products are

calcite:	$[Ca^{++}][CO_3^=]\gamma_{++}\gamma_= = 10^{-8.52}$,
brucite:	$[Mg^{++}][OH^-]^2\gamma_{++}\gamma_-^2 = 10^{-11.6}$,
magnesite:	$[Mg^{++}][CO_3^=]\gamma_{++}\gamma_= = 10^{-4.9}$,
nesquehonite:	$[Mg^{++}][CO_3^-]\gamma_{++}\gamma_= = 10^{-5.4}$,
hydromagnesite:	$[Mg^{++}]^4[CO_3^=]^3[OH^-]^2\gamma_+^4 \gamma_=^3\gamma_-^2 = 10^{-29.5}$,
lime:	$[Ca^{++}][OH^-]^2\gamma_{++}\gamma_-^2 = 10^{-5.3}$.

At each pH value, test to see which solubility products are satisfied and reduce C_{Ca} or C_{Mg} appropriately. Do *not* consider a model in which pH is shifted by precipitation; that's too complicated.

25. A geothermal power plant is to be operated on brine that is obtained from a well on the Reykjanes Peninsula, Iceland. The temperature is 221°C and the water composition is

(in mole/L):

$$[Na^+] = 10.44, \qquad C_T = 2.65$$
$$[K^+] = 1.382, \qquad [Cl^-] = 15.75$$
$$[Ca^{++}] = 1.812, \qquad [SiO_2] = 0.374$$
$$[Mg^{++}] = 0.008, \qquad [B(OH)_3] = 0.012$$
$$[SO_4^=] = 0.072, \qquad Fe_T = 0.048$$
$$[S^=] = 0.051,$$

Assuming that the only significant ion unaccounted for is $[H^+]$, estimate the partial pressure of CO_2 at 200°C by using data from Tables 2.2 and 2.4. Is the solution saturated with $CaCO_3$? What other analysis would you want most?

Answer: $P \cong 10^{+3.5}$; probably not; alkalinity or acidity

26. The ion-pair formation constant for $CaSO_4$ at 200°C has been estimated to be $10^{+3.6}$ (Helgeson, H. C. 1969. *Am. J. Sci.* 267:729–804). How does this information change the answer to Example 14 of this chapter?

Epilogue

Originally, I intended to have a Chapter 7, covering the biological applications of carbon dioxide and its equilibria, but two things persuaded me not to include it. First, most of the biochemical concern is with diffusion through membranes and with the enzyme-catalyzed reaction

$$CO_2 + H_2O \rightleftharpoons HCO_3^- + H^+,$$

which has, from an equilibrium point of view, already been adequately treated in Chapter 2. (For some up-to-date information on CO_2 research in the biochemical field, consult Bauer, C., Cros, G., and Bartels, H. eds., 1980. *Biophysics and Physiology of Carbon Dioxide.* Berlin: Springer-Verlag.)

The second factor was advice by the editors at Addison-Wesley that the audience for biochemical and medical books did not overlap the audience for geochemical and engineering books, so that if I was going to write about carbon dioxide for biochemists and medical students, that should be a different book from this one. As you can see, I took their advice.

Appendixes

Summary of
Important Results

CHAPTER 2 THE BASIC EQUATIONS

Equilibria:

$$[CO_2] = K_H P_{CO_2},$$ (2.1)

$$[H^+][HCO_3^-] = K_{a1}[CO_2],$$ (2.2)

$$[H^+][CO_3^=] = K_{a2}[HCO_3^-].$$ (2.3)

Charge balance:

$$A = [HCO_3^-] + 2[CO_3^=] + [OH^-] - [H^+] = \text{Carbonate alkalinity},$$ (2.16)

$$[H^+] + [Na^+] = [Cl^-] + [HCO_3^-] + 2[CO_3^=] + [OH^-],$$ (2.17)

$$A' = -A = \text{Mineral acidity}.$$

Mass balances:

$$[Na^+] = \text{Concentration of strong base (i.e., NaOH)},$$

$$[Cl^-] = \text{Concentration of strong acid (i.e., HCl)},$$

$$C_T = [CO_2] + [HCO_3^-] + [CO_3^=] + \cdots = \text{Total carbonate}.$$ (2.18)

Species concentrations at constant C_T (see also Fig. 2.4):

$$[CO_3^=] = C_T \frac{K_{a1}K_{a2}}{K_{a1}K_{a2} + K_{a1}[H^+] + [H^+]^2},$$ (2.21)

$$[HCO_3^-] = C_T \frac{K_{a1}[H^+]}{K_{a1}K_{a2} + K_{a1}[H^+] + [H^+]^2},$$ (2.22)

$$[CO_2] = C_T \frac{[H^+]^2}{K_{a1}K_{a2} + K_{a1}[H^+] + [H^+]^2}.$$ (2.23)

Alkalinity as a function of C_T and pH (Fig. 2.4):

$$A = C_T \frac{K_{a1}[H^+] + 2K_{a1}K_{a2}}{[H^+]^2 + K_{a1}[H^+] + K_{a1}K_{a2}} + \frac{K_w}{[H^+]} - [H^+] = C_T F - G.$$ (2.28) and (2.29)

231

Special points:

For pure CO_2 and H_2O, $A = 0$.

For $NaHCO_3$, $A = C_T$.

For Na_2CO_3, $A = 2C_T$.

Excess strong base is required for $A > 2C_T$.

Excess strong acid is required for $A < 0$.

A is invariant to addition or removal of CO_2.

A is a linear function of C_T at constant pH.

Effect of ionic strength:

$$I = \frac{1}{2} \sum_i C_i z_i^2. \tag{2.32}$$

Activity coefficient according to Davies equation:

$$\log \gamma_z = -0.5z^2 f(I),$$

$$f(I) = \left(\frac{I^{1/2}}{1 + I^{1/2}} - 0.2I \right) \left(\frac{298}{t + 273} \right)^{2/3}, \tag{2.36}$$

$$K_H^0 = \frac{[CO_2]\gamma_0}{P} = K_H \gamma_0, \tag{2.37}$$

$$\log \gamma_0 = +bI \quad (b = 0.04 \text{ to } 0.10, \text{ see Table 2.4}). \tag{2.38}$$

Equilibrium constants:

$$K_1' = \frac{10^{-\text{pH}}[HCO_3^-]}{[CO_2]} = K_{a1}\gamma_+ = K_{a1}^0 \frac{\gamma_0}{\gamma_-}, \tag{2.39}$$

$$K_2' = \frac{10^{-\text{pH}}[CO_3^=]}{[HCO_3^-]} = K_{a2}\gamma_+ = K_{a2}^0 \frac{\gamma_-}{\gamma_=}, \tag{2.42}$$

$$pK_1' = pK_{a1}^0 - 0.5\, f(I) - bI, \tag{2.40}$$

$$pK_2' = pK_{a2}^0 - 1.5\, f(I), \tag{2.43}$$

$$pK_w' = -\log(10^{-\text{pH}}[OH^-]) = pK_w^0 - 0.5\, f(I), \tag{2.48}$$

$$pK_{s0} = pK_{s0}^0 - 4.0\, f(I). \tag{2.51}$$

The values of pK_2', pK_w', and pK_{s0} for seawater are substantially less than given by Eqs. (2.43), (2.48), and (2.51) because of ion pairing (see Figs. 2.7, 2.8, and 2.9).

CHAPTER 3   THE ALKALINITY TITRATION CURVE

Alkalinity titration curve (see also Figs. 3.6 and 3.7):

$$V = V_0 \frac{A - C_T F + G}{C_a - G}, \tag{3.6}$$

where

$$F = \frac{K_{a1}[H^+] + 2K_{a1}K_{a2}}{K_{a1}K_{a2} + K_{a1}[H^+] + [H^+]^2} = \frac{K_{a1}}{K_{a1} + [H^+]} + \frac{K_{a2}}{K_{a2} + [H^+]} + \cdots \quad \text{(3.7) and (3.9)}$$

$$G = [H^+] - \frac{K_w}{[H^+]}. \tag{3.8}$$

Titration error at CO_2 endpoint $(V \cong AV_0/C_a)$:

$$E = \frac{CV - AV_0}{AV_0} = [H^+]\frac{A + C_a}{AC_a} - \frac{K_{a1}}{[H^+]}\frac{C_T}{A}, \tag{3.19}$$

$E = 0$ at equivalence point.

Relation between total alkalinity and carbonate alkalinity in the presence of weak acid (fully dissociated at the start of the titration) **and weak base** (fully protonated at the equivalence point):

$$A_T = A_c + [HA]' + [B]^0, \tag{3.34}$$

where $[HA]'$ is the concentration of undissociated acid at the equivalence point, $[B]^0$ the concentration of unprotonated base at the start of the titration.

Linearized (Gran) titration functions (Figs. 3.4 and 3.5):

$$\text{pH} < 5: \quad \text{plot } f_1 = (V + V_0)10^{-\text{pH}} \text{ versus } V \text{ to get } V_2 \text{ (CO_2 endpoint)}; \tag{3.39}$$
$$5 < \text{pH} < 8.3: \quad \text{plot } f_2 = (V_2 - V)10^{-\text{pH}} \text{ versus } V \text{ to get } V_1 \text{ (HCO_3^- endpoint)}; \tag{3.44}$$
$$\text{plot } f_3 = (V - V_1)10^{+\text{pH}} \text{ versus } V \text{ to get } V_2. \tag{3.45}$$

Buffer index at constant C_T:

$$\beta_c = \left(\frac{\partial A}{\partial \text{pH}}\right)_{C_T} = 2.303\frac{[HCO_3^-]}{C_T}([CO_3^=] + [CO_2]) + 2.303([H^+] + [OH^-]). \tag{3.62}$$

Buffer index at constant P_{CO_2}:

$$\beta_p = \left(\frac{\partial A}{\partial \text{pH}}\right)_P = 2.303([HCO_3^-] + 4[CO_3^=] + [OH^-] + [H^+]). \tag{3.66}$$

CHAPTER 4   SOLUBILITY EQUILIBRIA: CALCIUM CARBONATE

All the following equations hold in the presence of excess solid $CaCO_3$, except where noted.

Solubility product:

$$[Ca^{++}][CO_3^=] = K_{s0}. \tag{4.1}$$

Ion-pairing equilibria (at 25°C):

$$[CaHCO_3^+] = 10^{+1.22}[Ca^{++}][HCO_3^-]\frac{\gamma_{++}\gamma_-}{\gamma_+}, \tag{4.2}$$

$$[CaCO_3^0] = 10^{+3.1}[Ca^{++}][CO_3^=]\gamma_{++}\gamma_=, \tag{4.8}$$

$$[CaOH^+] = 10^{+1.3}[Ca^{++}][OH^-]\frac{\gamma_{++}\gamma_-}{\gamma_+}. \tag{4.9}$$

Charge balance for system containing CO_2, H_2O, $CaCO_3$ (solid), NaOH, and HCl:

$$[Na^+] + 2[Ca^{++}] + [CaOH^+] + [CaHCO_3^+] + [H^+]$$
$$= [HCO_3^-] + 2[CO_3^=] + [OH^-] + [Cl^-]. \tag{4.22}*$$

$CaCO_3$ and H_2O under partial pressure P of CO_2:

$$[H^+] = \left(\frac{K_{a1}^2 K_{a2} K_H^2 P^2}{2K_{s0}}\right)^{1/3} + \cdots. \qquad \text{see (4.12) and (4.19)}$$

Alkalinity:

$$A = \left(\frac{2K_{s0}K_{a1}K_H P}{K_{a2}}\right)^{1/3} + \cdots. \qquad \text{see (4.18) and (4.21)}$$

Addition of strong base or strong acid with constant $P^†$:

$$[Na^+] - [Cl^-] = \underbrace{\frac{K_{a1}K_H P}{[H^+]}\left(1 + \frac{2K_{a2}}{[H^+]}\right)}_{A} - \frac{2K_{s0}[H^+]^2}{K_{a1}K_{a2}K_H P} + \cdots, \tag{4.23) or (4.26}$$

$$[Na^+] - [Cl^-] = \qquad\qquad A \qquad\qquad - 2[Ca^{++}]. \tag{4.25}$$

Saturation point where $CaCO_3$ just barely dissolves or precipitates:

$$[H^+] = \left(\frac{C_{Ca}K_{a1}K_{a2}K_H P}{K_{s0}}\right)^{1/2}. \tag{4.30}$$

Amount of strong acid or base needed to reach saturation point:

$$[Cl^-] - [Na^+] = 2C_{Ca} - \left(\frac{K_{s0}K_{a1}K_H P}{C_{Ca}K_{a2}}\right)^{1/2}\left[1 + \left(\frac{4K_{s0}K_{a2}}{C_{Ca}K_{a1}K_H P}\right)^{1/2}\right] + \cdots. \tag{4.31}$$

In unsaturated region‡:

$$[Cl^-] - [Na^+] = 2C_{Ca} - \frac{K_{a1}K_H P}{[H^+]}\left(1 + \frac{2K_{a2}}{[H^+]}\right) + [H^+] + \cdots. \tag{4.29}$$

Solution saturated with $CaCO_3$ but without gas phase:

$$[H^+] = \frac{C_T^2 K_{a2}}{2K_{s0}} + \cdots \qquad (\text{pH} \cong 8) \tag{4.38}$$

---

* This is (4.10) if $[Na^+] = [Cl^-] = 0$.

† The term is parentheses should contain the additional terms $-[H^+] + K_w/[H^+]$, which are usually neglected.

‡ Equation (4.29) should contain the additional term, $-K_w/[H^+]$, which is usually neglected.

Invariants with respect to precipitation or dissolution of $CaCO_3$:

$$D = 2(C_T - [Ca^{++}]), \tag{4.40}$$

$$D_2 = D - [Na^+] + [Cl^-] = [HCO_3^-] + 2[CO_2] + [H^+] - [OH^-] + \cdots . \tag{4.42}$$

General solution when acid or base is added to a saturated solution:

$$[Na^+] - [Cl^-] = D - x - [H^+] + \frac{K_w}{[H^+]}, \tag{4.48}$$

where

$$x = \frac{z}{4y} \left[D + \left(D^2 + \frac{16 K_{s0}}{K_{a1} K_{a2}} y \right)^{1/2} \right], \tag{4.53}$$

$$y = K_{a1} K_{a2} + K_{a1}[H^+] + [H^+]^2, \tag{4.49}$$

$$z = K_{a1}[H^+] + 2[H^+]^2. \tag{4.50}$$

Approximate solution for pH $\cong$ 8:

$$[H^+] = \frac{K_{a2}}{2 K_{s0}} (D - [Na^+] + [Cl^-])(D - 2[Na^+] + 2[Cl^-]) + \cdots$$

$$= \frac{K_{a2}}{2 K_{s0}} (D_2)(2 D_2 - D) + \cdots . \tag{4.47}$$

Saturation point:

$$\frac{C_{Ca} C_T K_{a2}}{K_{s0}} = K_{a2} + [H^+] + \frac{[H^+]^2}{K_{a1}} = [H^+] + \cdots . \tag{4.55}\text{ or }(4.56)$$

Unsaturated region:

$$[Cl^-] - [Na^+] = 2 C_{Ca} - C_T \frac{K_{a1}[H^+] + 2 K_{a1} K_{a2}}{K_{a1} K_{a2} + K_{a1}[H^+] + [H^+]^2} - \frac{K_w}{[H^+]} + [H^+]$$

$$= 2 C_{Ca} - C_T \frac{K_{a1}}{K_{a1} + [H^+]} + [H^+] + \cdots . \tag{4.58}$$

Buffer index at constant partial pressure in the presence of $CaCO_3$:

$$\beta_{P,s} = \beta_P + 9.21[Ca^{++}] + \cdots , \tag{4.65}$$

where

$$\beta_P = 2.303([HCO_3^-] + 4[CO_3^=] + [OH^-] + [H^+]). \tag{3.66}$$

Buffer index at constant C_T in the presence of $CaCO_3$:

$$\beta_{C,s} = \beta_C + 4.61[Ca^{++}] \frac{[HCO_3^-] + 2[CO_2]}{C_T} + \cdots , \tag{4.67}$$

where

$$\beta_C = 2.303 \left(\frac{[HCO_3^-]([CO_3^=] + [CO_2])}{C_T} + [OH^-] + [H^+] \right). \tag{3.62}$$

Buffer index at constant D in the presence of $CaCO_3$:

$$\beta_{D,s} = \beta_C + 4.61[Ca^{++}] \frac{([HCO_3^-] + 2[CO_2])^2}{DC_T} + \cdots \qquad (4.71)$$

CHAPTER 5   APPLICATIONS TO GEOCHEMISTRY AND OCEANOGRAPHY

Rain water ($CO_2 + H_2O$):

$$pH = \frac{1}{2}(pK_{a1}^0 + pK_H^0 - \log P) + \cdots. \qquad (5.1)$$

Acid rain (NO_3^-, $SO_4^=$, H^+, etc.; see Table 5.1):

$$A' = \frac{10^{-pH}}{\gamma_+} - \frac{10^{+pH}K_{a1}^0 K_H^0 P}{\gamma_-} + \cdots. \qquad (5.6)$$

River water (see Tables 5.2 and 5.3):
$CaCO_3 + CO_2 + HX + H_2O$ (unsaturated):

$$2[Ca^{++}] - [X^-] = 2(C_T - G) - [X^-] = A = -A', \qquad \textbf{(5.9) and (5.12)*}$$

$$C_T = (2G + [X^-]) \frac{K_1' + 10^{-pH}}{K_1' + 2 \cdot 10^{-pH}} + \cdots \qquad (5.15)$$

$CaCO_3 + CO_2 + H_2O$ (saturated):

$$pH = \frac{1}{3}(pK_{a2}^0 - pK_{s0}^0) + \frac{2}{3}(pK_{a1}^0 + pK_H^0 - \log P) + \frac{1}{3}\log 2 + 0.5 f(I) + \cdots, \qquad \textbf{(5.16)}$$

$$\log A = \frac{1}{3}(pK_{a2}^0 - pK_{a1}^0 - pK_H^0 - pK_{s0}^0 + \log P + \log 2) + f(I) + \cdots. \qquad (5.17)$$

Seawater (see Tables 5.4 and 5.5):

Density of $35‰$ seawater at $25°C = 1.0235$ g/cm^3

$(‰$ salinity$) = 0.03 + 1.805(‰$ chlorinity$)$, $\qquad$ (5.21)

Ionic strength $= 0.716$ mole/L for standard $35‰$,

$$\gamma_+ = 0.69, \qquad \gamma_{++} = 0.23.$$

Ion pairs (see Tables 5.6 and 5.7), for example:

$$[Mg^{++}]_T = [Mg^{++}] + [MgSO_4] + [MgCO_3] + [MgHCO_3^+] + \cdots$$

---

* Activity coefficients of ions are given by the Davies equation (2.36) and activity coefficients of CO_2 by equation (5.38) with $b = 0.1$:

$$f(I) = \left(\frac{I^{1/2}}{1 + I^{1/2}} - 0.2I\right)\left(\frac{298}{t + 273}\right)^{2/3}.$$

Total alkalinity versus carbonate alkalinity:

$$A_c = [HCO_3^-]_T + 2[CO_3^=]_T + [OH^-] - [H^+], \qquad \text{(3.1) or (5.34)}$$

$$A_T = A_c + \sum_i [HA]_i' + \sum_j [B]_j^0. \qquad \text{(3.35) or (5.33)}$$

Superscript zero denotes the initial solution (pH $\cong$ 8), prime denotes the solution at the alkalinity titration endpoint (pH $\cong$ 4.5). For seawater, 99.2% of A_T is due to carbonate and borate. Other species may contribute small (and sometimes variable) amounts (percentages indicated under the equation terms):

$$A_T = \underset{96.3\%}{A_c} + \underset{2.9\%}{[B(OH)_4^-]^0} + \underset{0.2\%}{[SiO(OH)_3^-]^0} + \underset{<0.1\%}{[HPO_4^=]^0} + \underset{<0.05\%}{[PO_4^\equiv]^0}$$

$$+ \underset{0.1\%}{[MgOH^+]^0} + \underset{0.4\%}{[HSO_4^-]'} + \underset{0.02\%}{[HF]'} + \underset{<0.001\%}{[H_3PO_4]'}$$

$$+ [NH_3]^0 + [HS^-]^0 + 2[S^=]^0.$$
$$\text{(anoxic waters only)}$$

Field observations on seawater:

Figure 5.3: $A_T = 2.25 \cdot 10^{-3}$ mole/kg at surface, $2.45 \cdot 10^{-3}$ mole/kg at 4000 m depth.

Figure 5.4: $C_T = 2.0 \cdot 10^{-3}$ mole/kg at surface, 2.3 to $2.4 \cdot 10^{-3}$ mole/kg at 4000 m.

Figure 5.5: $P_{CO_2} = 3 \cdot 10^{-4}$ atm at surface, $1.2 \cdot 10^{-3}$ atm at 500 m, $6 \cdot 10^{-4}$ atm at 4000 m.

Figure 5.6: $2C_T - A = 1.75 \cdot 10^{-3}$ at surface, $2.4 \cdot 10^{-3}$ at 1000 m, $2.25 \cdot 10^{-3}$ at 4000 m.

Most surface seawater is supersaturated with calcite, most deep waters (> 3000 m) are undersaturated (Fig. 5.10). This results from changes in pressure and temperature (Figs. 5.8 and 5.9).

Increasing atmospheric CO_2 (Figs. 5.11 and 5.12) is partly absorbed by the upper mixed layer of the ocean. Revelle buffer factor:

$$B = \left(\frac{\partial \log P}{\partial \log C_T}\right)_A = \frac{C_T}{[CO_2] + [CO_3^=]} + \cdots. \qquad (5.48)$$

Approximately half the current increase in P_{CO_2} is being absorbed by the ocean.

Mixing fresh and estuarine waters is normally modeled by making salinity (or chlorinity) conservative:

$$Cl - Cl^r = b(Cl^0 - Cl^r), \qquad (5.51)$$

where Cl is the chlorinity of the mixed water, Cl^r that of the river, Cl^0 that of the ocean, and b the fraction of the water that originated in the ocean. Other conservative parameters are assumed to be linear in b:

$$A = bA^0 + (1 - b)A^r. \qquad (5.53)$$

This would not hold if $CaCO_3$ was dissolving or precipitating as the waters were mixed; pH follows from A and C_T via Eqs. (5.55) to (5.57).

Mixing two waters saturated in CaCO$_3$ may yield a saturated, unsaturated, or supersaturated solution depending on the temperatures and the presence of other ionic components (Figs. 5.17 and 5.18).

Evaporation of natural waters produces brines of varying composition (see Fig. 5.19). The minerals that crystallize first are usually calcite (CaCO$_3$) and related minerals containing Mg substituted for Ca. (The highly ordered dolomite structure with composition CaMg(CO$_3$)$_2$ is usually found only in older rocks.) Sulfate-rich waters produce gypsum CaSO$_4 \cdot 2$H$_2$O and related minerals (see for example Table 5.9). Bicarbonate-rich waters produce alkaline salts such as trona (NaHCO$_3 \cdot$ Na$_2$CO$_3 \cdot 2$H$_2$O) (see Fig. 5.20).

Reactions at high temperatures and pressures (hydrothermal conditions) differ from those at ordinary environmental conditions primarily in their different values of thermodynamic constants (compare Fig. 5.27 and Eq. (5.17)). These are often measured under the conditions of interest, but can be sometimes estimated by standard thermodynamic methods (Eqs. (5.58)–(5.63)). For example (see also Table 2.2):

Figure 5.24: pK_H^0 varies from 1.5 at 25°C to 2.1 at 150°C to 1.7 at 330°C;

Figure 5.26(a): pK_{a1}^0 varies from 6.3 at 25°C to 8.7 at 330°C;

Figure 5.26(b): pK_{a2}^0 varies from 10.3 at 25°C to 10.1 at 100°C to 11.8 at 330°C;

Figure 5.27(b): pK_{s0}^0 varies from 8.3 at 25°C to 11.6 at 200°C.

Ionic-strength effects at higher temperatures and pressures are similar to room temperature and pressure. The coefficient b (log K_H/I, given in Table 2.4) increases from 0.1 at 25°C to 0.2 at 330°C. The Davies equation (2.36) will be less accurate at high temperatures (because the temperature dependence of the dielectric constant of water is not taken into account), but should give serviceable estimates. For several salts, $\gamma_\pm$ varies less than 20% from 0 to 100°C (Fig. 5.29(a)); for NaCl $\gamma_\pm$ varies less than 15% (at $I = 5$) as pressure is increased to 2000 atm (Fig. 5.29(b)). Calcite solubility increases with increasing P_{CO_2}, decreases with increasing temperature at constant pressure, and increases with addition of salt (Figs. 5.27, 5.30(a), 5.30(b)). At high concentrations of CO$_3^=$ and temperatures from 200 to 400°C, calcite solubility increases with [CO$_3^=$]^3, instead of decreasing inversely with [CO$_3^=$]; this implies the existence of complexes such as CaCO$_3^0$, Ca(CO$_3$)$_2^=$, Ca(CO$_3$)$_3^{4-}$, and Ca(CO$_3$)$_4^{6-}$ (see Fig. 5.31 and Problem 63).

CHAPTER 6   ENGINEERING APPLICATIONS: WATER CONDITIONING

Ionic strength from incomplete analyses:

Residual TDS = Total dissolved solids $-\sum$ All known solutes.

$$R = \text{TDS} - \sum_i 10^3 c_i M_i,$$

$$I = \frac{1}{2}\sum c_i z_i^2 + 2.50 \cdot 10^{-5} R$$

(species analyzed has concentration c_i mole/L, charge z_i, molecular weight M_i; R and TDS are in mg/L).

Addition of acid or base follows the equations of alkalinity with C_T invariant in a closed system, e.g.:

$$A = C_T F - G, \qquad\qquad \text{(2.30) or (6.5)}$$

$$F = \frac{K_{a1}[H^+] + 2K_{a1}K_{a2}}{[H^+]^2 + K_{a1}[H^+] + K_{a1}K_{a2}}$$

$$= \frac{K_{a1}^0 10^{-pH}(\gamma_0/\gamma_-) + 2K_{a1}^0 K_{a2}^0/\gamma_=}{10^{-2pH} + K_{a1}^0 10^{-pH}(\gamma_0/\gamma_1) + K_{a1}^0 K_{a2}^0/\gamma_=}, \qquad\qquad \text{(6.6) or (6.9)}$$

$$G = [H^+] - K_w/[H^+] = \frac{10^{-pH}}{\gamma_+} - \frac{K_w^0 10^{+pH}}{\gamma_-}. \qquad\qquad \text{(6.7)}$$

Addition or removal of CO_2: A is invariant, C_T changes. System is normally at constant P_{CO_2}.

Addition of lime ($Ca(OH)_2$): C_T is invariant, both A and $[Ca^{++}]$ increase. $\Delta A = 2x$, $\Delta[Ca^{++}] = x$, where x is the amount added.

Addition of limestone ($CaCO_3$): $K_{s0} = [Ca^{++}][CO_3^=]$ holds. D and D_2 are invariant (see (4.40) and (4.42)).

Langelier index of $CaCO_3$ saturation:

$$L = pH_{obs} - pH_{sat} = pH_{obs} - (pK_2' - pK_{s0}) + \log A + \log[Ca^{++}] + \cdots, \qquad \text{(6.32)}$$

$$L < 0 \text{ undersaturated,} \qquad L > 0 \text{ supersaturated.}$$

To adjust for ionic strength:

$$pK_2' - pK_{s0} = pK_{a2}^0 - pK_{s0}^0 - 2.5f(I), \qquad\qquad \text{(6.33)}$$

$$f(I) = \left(\frac{I^{1/2}}{1 + I^{1/2}} - 0.2I\right)\left(\frac{298}{273 + t}\right)^{2/3}. \qquad\qquad \text{(2.36)}$$

Lime–soda water softening process: If Ca^{++} and Mg^{++} are in excess over HCO_3^-, stoichiometry leads to the following conventions:

$$Ca(OH)_2 \text{ added:} \quad Q_1 = [CO_2]^0 + [Mg^{++}]^0 + \frac{1}{2}[HCO_3^-]^0, \qquad \text{(6.43)}$$

$$Na_2CO_3 \text{ added:} \quad Q_2 = [Ca^{++}]^0 + [Mg^{++}]^0 - \frac{1}{2}[HCO_3^-]^0. \qquad \text{(6.44)}$$

If HCO_3^- is in excess over Ca^{++} and Mg^{++}, then

$$Q_1 = [CO_2]^0 + [HCO_3^-]^0 - [Ca^{++}]^0, \qquad\qquad \text{(6.45)}$$

$$Q_2 = 0. \qquad\qquad \text{(6.46)}$$

If HCO_3^- is in excess over Ca^{++} but not over both Mg^{++} and Ca^{++}, see (6.43) and (6.44):

Noncarbonate hardness $= 2[Ca^{++}]^0 + 2[Mg^{++}]^0 - A = 2Q_2 + \cdots$.

If concentrations Q_1 and Q_2 are added according to stoichiometry (6.43) through (6.46), then

$$C_T = K_{s0}^{1/2} + \cdots, \tag{6.62}$$

$$pH = \frac{1}{2} \log\left(\frac{K_{s0}^{1/2}}{K_2' K_w'} + 2K_{s0}^M \frac{10^{-pH}}{K_w'^3}\right) + \cdots, \tag{6.65}$$

which is solved iteratively with a starting value $pH = 10.6$.

Residual hardness $= [Ca^{++}] + [Mg^{++}]$

$$= K_{s0}^{1/2}\left(1 + \frac{10^{-pH}}{K_2'}\right) + \frac{K_{s0}^M 10^{-2pH}}{K_w'^2}. \tag{6.67}$$

Useful Data Tables

Table 2.1

Concentration equilibrium constants at 25°C and different ionic strengths*

Constant	Ionic strength						
	0	0.1	0.2	0.7	1.0	1.0	3.5
pK_H	1.47	1.48 (SW)	1.49 (SW)	1.54 (SW)		1.51 (NaClO$_4$)	1.55 (NaClO$_4$)
pK_{a1}	6.35	6.15 (KCl)	6.06 (0.26 KNO$_3$)	5.86 (SW)	5.99 (KNO$_3$)	6.04 (NaClO$_4$)	6.33 (NaClO$_4$)
pK_{a2}	10.33	9.92 (NaCl)	9.82 (NaCl)	8.95 (SW)	9.37 (NaCl)	9.57 (NaClO$_4$)	9.56 (NaClO$_4$)
pK_w	13.999	13.78 (NaClO$_4$)	13.73 (0.26 KNO$_3$)	13.20 (SW)	13.74 (NaCl)	13.78–13.86 (NaClO$_4$)	14.42 (3.8 M NaClO$_4$)
pK_{s0} (calcite)	8.34–8.52	7.16 (SW)	6.87 (SW)	6.34 (SW)		6.93 (NaClO$_4$)	7.18 (NaClO$_4$)

* $pK_H = -\log K_H$, $pK_{a1} = -\log K_{a1}$, $pK_{a2} = -\log K_{a2}$, $pK_{s0} = -\log K_{s0}$. See footnote to Table 2.2 regarding pK_{s0}^0 and ion-pairing effects. SW—seawater of appropriate salinity, i.e., $I = 0.7$ (35‰), taken from Table 5.5 (Hansson scale). See also data in Figs. 2.6 to 2.9. Data selected from Sillén, L. G. and Martell, A. E. *Stability Constants*. Special publ. no. 17 (1964) and 25 (1971). London: The Chemical Society; Riley, J. P. and Skirrow, G. 1975. *Chemical Oceanography*, vol. 2. New York: Academic Press, pp. 173–180, Butler, J. N. and Huston, R. 1975. *J. Phys. Chem.* 74:2976–2983; Edmond, J. M. and Gieskes, J. M. T. M. 1970. *Geochem. Cosmochim. Acta* 34: 1261–1291.

Table 2.2
Selected values of equilibrium constants, $0°$ to $50°C$ (extrapolated to zero ionic strength)*

Temperature (°C)	pK_H^0 (mole/L·atm)	pK_{a1}^0 (mole/L)	pK_{a2}^0 (mole/L)	pK_w^0 (mole/L)^2	pK_{s0}^0 (Calcite) (mole/L)^{2†}	pK_{s0}^0 (corrected for ion pair)†
0	1.11	6.579	10.625	14.955	8.03	8.37
5	1.19	6.517	10.557	14.734	8.09	8.39
10	1.27	6.464	10.490	14.534	8.15	8.41
15	1.33	6.419	10.430	14.337	8.22	8.45
20	1.41	6.381	10.377	14.161	8.28	8.48
25	1.47	6.352	10.329	13.999	8.34	8.52
30	1.53	6.327	10.290	13.833	8.40	8.57
35	1.59	6.309	10.250	13.676	8.46	8.62
40	1.64	6.298	10.220	13.533	8.51	8.66
45	1.68	6.290	10.195	13.394	8.56	8.71
50	1.72	6.285	10.172	13.263	8.62	8.76
100	1.99	6.45	10.16	12.27	9.62	
150	2.07	6.73	10.33	11.64	10.54	
200	2.05	7.08	10.71	11.28	11.62	

* pK_H^0, pK_{a1}^0, pK_{a2}^0, pK_{s0}^0 were selected by Stumm, W. and Morgan, J. J. 1981. *Aquatic Chemistry*. Ed. 2 New York: Wiley, pp. 204–206. Some interpolations have been made to fill out the table at $5°$ intervals. Full bibliography and actual literature values (including pK_w^0) are found in Sillén, L. G. and Martell, A. E. *Stability Constants*. Special publ. no. 17 (1964) and 25 (1971). London: The Chemical Society.

† The first column gives the traditional solubility product, extrapolated from measurements at low ionic strength. However, the presence of ion pairs, particularly $CaHCO_3^+$, biases the extrapolation and leads to a pK_{s0}^0 that is too low. When corrections for this ion pair are made, the values in the second column are obtained. For accurate calculations, the equilibrium $[CaHCO_3^+] = 10^{+1.23}[Ca^{++}]$ $[HCO_3^-]$ should be included (see Chapters 4 and 5). (Christ, C. L., Hostetler, P. B., and Siebert, R. M. 1974. J. Res. US Geol. Survey 2:175–184.)

Table 5.1
Average composition of rainfall in northeastern United States (micromoles/liter)*

$[SO_4^=] = 28.1 \pm 3.9$	$[Cl^-] = 12.1 \pm 12.7$
$[NO_3^-] = 25.9 \pm 5.0$	$[Na^+] = 9.3 \pm 12.2$
$[NH_4^+] = 16.0 \pm 4.6$	$[H^+] = 72.3 \pm 12.3$
	pH = 4.14 ± 0.07

* Pack, D. H. 1980. Precipitation chemistry patterns: a two-network data set. *Science* 208:1143–1145. These data are from the MAP3S Network.

Table 5.2
Average composition of river water (10^{-5} mole/L)*

Component	N. America	S. America	Europe	Asia	Africa	Australia	World
HCO_3^-	111	51	156	129	70	52	95.7
$SO_4^=$	21	5.0	25	8.7	14	2.7	11.7
Cl^-	23	14	19	25	34	28	22.0
NO_3^-	1.7	1.2	6.3	1.2	1.4	0.1	1.7
F^-	0.8						0.5
Ca^{++}	53	18	78	46	31	9.8	37.5
Mg^{++}	21	6.2	23	23	16	11	16.9
Na^+	39	17	23	40	48	13	27.4
K^+	3.6	5.1	4.3			3.6	5.9
Fe	0.29	2.5	1.4	0.02	2.3	0.53	1.2
SiO_2	15	20	12	20	39	6.5	21.8
Al	0.9?						<1.5?
Anionic charge	178	76	231	173	133	86	143.3
Cationic charge	191	71	229	178	142	58	142.1
Charge balance	+13	−5	−2	+5	+9	−28	−1.2
Ionic strength	280	102	356	253	199	95	209
$\dfrac{2[Ca^{++}]}{[HCO_3^-]}$	0.95	0.71	1.00	0.71	0.89	0.38	0.78
$-\log P_{CO_2}$ for $CaCO_3$-saturation	3.32	4.41	2.88	3.25	3.93	4.65	3.58

* These data were compiled by Livingstone, D. A. (1963. *Chemical Composition of Rivers and Lakes.* US Geological Survey Professional Paper 440G, Washington DC: US Government Printing Office.). These same data are also quoted by Stumm and Morgan (1970, p. 385) and Holland (op. cit., p. 93) (note errors in the molar value for Mg^{++} in both books). I have converted Livingstone's ppm values to mole/kg (mole/L within roundoff errors) by using atomic weights based on $^{12}C = 12.0000$; Fe and Al are included with silica as uncharged components, because in the normal pH range of 6 to 8 both Fe(III) and Al(III) are probably present principally as colloidal particles. Data for Al and F^- are from Holland (loc. cit.), who also gives a review of other data on trace element concentrations in river water (op. cit., pp. 136–138). Another recent review is by Martin, J. M., and Meybeck, M., 1979. *Mar. Chem.* 7:173–206; see p. 116 for calculation of last line.

Table 5.4
Composition of normal seawater* and world average
river water† (mmole/kg)

Component	Seawater	River water
Na^+	468.04	0.274
K^+	10.00	0.059
Mg^{++}	53.27	0.169
Ca^{++}	10.33	0.375
Sr^{++}	0.10	
Cl^-	545.88	0.220
$SO_4^=$	28.20	0.117
Br^-	0.83	
F^-	0.07	0.0053
$HCO_3^- \ (+CO_2 + CO_3^=)$	2.2–2.5‡	0.957
$B(OH)_3 + B(OH)_4^-$	0.43	
$Si(OH)_4 + SiO(OH)_3^-$	0.001–0.1‡	0.218
$H_2PO_4^- + HPO_4^= + PO_4^{\equiv}$	0.0001–0.005‡	
NO_3^-	0.0001–0.05	0.017
pH	(7.4 to 8.3)	(6.0 to 8.5)
Ionic strength	700	2.09

* After Hansson, I. 1973. *Deep Sea Res.* 20:479–491.

† See Table 5.2. The relationship between river water composition and seawater composition is beyond the scope of this book, but is discussed in detail by Holland (op. cit. Chapter 5) in terms of mass balances. This approach has largely supplanted the elegantly simple equilibrium theory of seawater composition put forth by L. G. Sillén (1961 "The Physical Chemistry of Sea Water," in *Oceanography*, M. Sears, ed. Washington, D.C.: Amer. Assoc. Advancement Sci. Publ. No. 67. pp. 549–581) and summarized by Stumm and Morgan (op. cit. pp. 572–574). See Problem 53. Another approach to the global mass balance of elements will be found in these papers: Whitfield, M., 1979. *Mar. Chem.* 8:101–123; Whitfield, M., and Turner, D. R., 1979. *Nature* (London) 278:132–137; Turner, D. R., Dickson, A. G., and Whitfield, M., 1980. *Mar. Chem.* 9:211–218.

‡ Biologically active species vary considerably with time and sampling location.

Table 5.5
Hybrid equilibrium constants for sea water at 25°C, 35‰ salinity*

$[H^+][OH^-] = 10^{-13.20}$ (K_w, mole²/L²)

$10^{-pH}[OH^-] = 10^{-13.60}$ (K'_w, mole/L, NBS pH scale)

$[CO_2] = 10^{-1.536}P_{CO_2}$ (K_H, mole/L · atm)

$[CO_2] = 10^{-1.547}P_{CO_2}$ (mole/kg · atm)

$10^{-pH}[HCO_3^-] = 10^{-6.00}[CO_2]$ (K'_1, NBS pH scale)

$[H^+][HCO_3^-] = 10^{-5.857}[CO_2]$ (K_{a1}, mole/kg, Hansson scale)

$10^{-pH}[CO_3^=] = 10^{-9.12}[HCO_3^-]$ (K'_2, NBS pH scale)

$[H^+][CO_3^=] = 10^{-8.947}[HCO_3^-]$ (K_{a2}, mole/kg, Hansson scale)

$10^{-pH}[B(OH)_4^-] = 10^{8.71}[B(OH)_3]$ (K'_B, NBS pH scale)

$[H^+][B(OH)_4^-] = 10^{-8.61}[B(OH)_3]$ (K_B, mole/kg, Hansson scale)

$10^{-pH}[SiO(OH)_3^-] = 10^{-9.4}[Si(OH)_4]$ (K'_{Si})

$10^{-pH}[H_2PO_4^-] = 10^{-1.61}[H_3PO_4]$ (K_1^P)

$10^{-pH}[HPO_4^=] = 10^{-6.08}[H_2PO_4^-]$ (K_2^P)

$10^{-pH}[PO_4^\equiv] = 10^{-8.56}[HPO_4^=]$ (K_3^P)

$10^{-pH}[SO_4^=] = 10^{-1.1}[HSO_4^-]$ (K'_a)

$[Ca^{++}][CO_3^=] = 10^{-6.34}$ ($K_{s0}^{calcite}$, mole²/kg²)

$[Ca^{++}][CO_3^=] = 10^{-6.13}$ ($K_{s0}^{aragonite}$, mole²/kg²)

---

* Most data are from Riley and Skirrow (op. cit.): Table 9.16 (after S. E. Ingle et al., 1973. *Mar. Chem.* 1:295–307); Table A9.9 (after C. Mehrbach et al., 1973. *Limnol. Oceanogr.* 18:897–907); Table A9.5 (after J. Lyman, 1956. Ph. D. Thesis, University of California at Los Angeles); Tables A9.6, A9.7, and A9.8 (after I. Hansson, 1973. *Deep-Sea Research* 20:461–491). Silicate constant estimated from 0.5 M NaCl data by J. M. T. M. Gieskes (1974. Chapter 3 in *The Sea*, E. Goldberg, ed. New York: Wiley). Phosphorus data from Kester, D. R., and Pytkowicz, R. M. (1967. *Limnol. Oceanogr.* 12:243–252). Sulfate data from Culberson, C. H., Pytkowicz, R. M., and Hawley, J. E. (1970. *J. Mar. Res.* 28:15–21).

Note that although pH depends on the scale used for calibration, 10^{-pH} does not, strictly speaking, carry any concentration units. On the other hand, since Hansson's "pH" scale is referred to a hydrogen ion concentration in moles/kg of seawater, his constants have been given the units mole/kg. The concentration units of conjugate acid and base (e.g. HCO_3^- and $CO_3^=$) cancel, and so the equilibrium constant has the same value whether these are expressed as moles/L or moles/kg. Where the units make a difference (as in K_H or K_{S0}) conversion can be made using the density of seawater: 1.0235 kg/L.

Note also that ionic concentrations include ion pairs with the major seawater ions of opposite charge, e.g., $[Ca^{++}]$ means $[Ca^{++}] + [CaCl^+] + [CaSO_4] + \cdots$ and $[CO_3^=]$ means $[CO_3^=] + [NaCO_3^-] + [MgCO_3] + \cdots$.

Table 5.7
Composition of typical ion-pair seawater model (mmol/kg)

$[Na^+] = 469.2$	$[K^+] = 9.8$	$[Mg^{++}] = 46.3$	$[Ca^{++}] = 9.2$	$[Cl^-] = 554.3$
$[NaCl] = 0$ (by assumption)				

$[Cl^-] = 554.3$					$[Cl^-]_T = 554.3$
$[SO_4^=] = 14.2$	$[NaSO_4^-] = 5.7$	$[KSO_4^-] = 0.2$	$[MgSO_4] = 7.2$	$[CaSO_4] = 1.1$	$[SO_4^=]_T = 28.4$
$[CO_3^=] = 0.03$	$[NaCO_3^-] = 0.05$		$[MgCO_3] = 0.2$	$[CaCO_3] = 0.02$	$[CO_3^=]_T = 0.3$
$[HCO_3^-] = 1.7$	$[NaHCO_3] = 0.2$		$[MgHCO_3^+] = 0.4$	$[CaHCO_3^+] = 0.08$	$[HCO_3^-]_T = 2.4$
	$[Na^+]_T = 475.2$	$[K^+]_T = 10.0$	$[Mg^{++}]_T = 54.1$	$[Ca^{++}]_T = 10.4$	

$$I_{tot} = 707 \qquad I_{i.p.} = 660$$

This model is consistent with the following formation constants for ion pairs:

$$K = \frac{[MX]}{[M][X]}$$

Species	log K (seawater)	log K^{0*}		Species	log K (seawater)	log K^{0*}
$NaSO_4^-$	-0.1	$+0.72$		$MgHCO_3^+$	$+0.7$	$+1.16$
$NaCO_3^-$	$+0.6$	$+1.27$		$MgOH^+$	$+1.9$	$+2.58$
$NaHCO_3$	-0.6	-0.25		$CaSO_4$	$+0.9$	$+2.31$
KSO_4^-	$+0.2$	$+0.96$		$CaCO_3$	$+1.9$	$+3.2$
$MgSO_4$	$+1.0$	$+2.36$		$CaHCO_3^+$	$+0.7$	$+1.26$
$MgCO_3$	$+2.1$	$+3.4$		$CaOH^+$	$+0.6$	$+1.30$

* From Garrels and Christ (op. cit., p. 96). See also Problem 56, p. 182, for a more recent choice of constants.

Annotated Bibliography

Butler, James N. 1964. *Ionic Equilibrium*. Reading, Mass.: Addison-Wesley.
Many detailed examples of acid-base, solubility, and complex-formation problems. The carbonate system is given brief attention as an example of a diprotic acid and its salts.

Butler, James N. 1964. *Solubility and pH Calculations*. Reading, Mass.: Addison-Wesley.
A 100-page extract from the more elementary parts of *Ionic Equilibrium*. Detailed examples of monoprotic acids and solubility product calculations expressed in concentration equilibrium constants, but almost no mention of the carbon dioxide system. (That's why this book was eventually written!)

Fair, Gordon M., Geyer, John C., and Okun, Daniel A. 1968. *Water and Wastewater Engineering*. (2 volumes) New York: Wiley.
A comprehensive treatise and textbook, of which Chapters 24 through 30 are most relevant to someone interested in water chemistry. Section 29-6 is the basis of the discussion in this book on the stoichiometry of the lime–soda process.

Garrels, Robert M. and Christ, Charles L. 1965. *Solutions, Minerals, and Equilibria*. New York: Harper & Row.
The second version of a pioneering monograph on the use of thermodynamics in geochemistry. Here you will find a lot about activity coefficients and ion pairs, as well as numerous phase diagrams for systems of geochemical importance. The five cases described for $CaCO_3$ solubility calculations (pure water, P_{CO_2} fixed, C_T fixed, pH fixed, and "water originally open to atmospheric CO_2, then closed before addition of $CaCO_3$") will be very familiar to you if you have read Chapters 4 and 5 of this book.

Goldberg, E. D. *The Sea* (6 volumes). New York: Wiley.
See especially Volume 5 (1974), Chapter 3, "The Alkalinity–Total CO_2 System in Seawater" by J. M. Gieskes and Chapter 2, "The Effects of Pressure on Dissociation Constants . . ." by A. Disteche. The material covered by these two chapters is similar to that in Riley and Skirrow, Volume 2, Chapter 9, but has its own viewpoint.

Holland, Heinrich D. 1978. *The Chemistry of the Atmosphere and Oceans*. New York: Wiley.
The theme of this book is that natural waters are regulated in their composition by geochemical weathering and deposition cycles. Its treatment of the chemical composition of rivers and oceans is especially thorough and takes up most of the book. A considerable part of Chapter 5 was inspired by this book.

Lowenthal, R. E. and Marais, G. vR. 1976. *Carbonate Chemistry of Aquatic Systems: Theory and Application*. Ann Arbor, Mich.: Ann Arbor Science Publishers.

Although my style and notation are different from this book, many of the same topics are covered. Emphasis is on calculations such as I described in the examples of Chapter 6: addition of acid and base to industrial and domestic water, and adjustment of the degree of saturation with $CaCO_3$, removal of Ca and Mg with lime and soda. A great deal of space is devoted to the Caldwell-Lawrence diagrams and related graphical methods for doing calculations relating alkalinity, calcium ion concentration, and pH.

Martell, Arthur E. and R. M. Smith, eds. *Critical Stability Constants*. (4 volumes) 1974–1977. New York: Plenum.
These volumes are based primarily on the Sillén and Martell compilation, but with some updating. They are a lot easier for a novice to read, and have the advantage of presenting you with an authoritative choice of constants when sometimes that choice is complicated and difficult. The disadvantage is that you do not easily have a full bibliography if you want to make your own critical choices.

Pytkowicz, R. M., ed. 1979. *Activity Coefficients in Electrolyte Solutions*. (2 volumes) Boca Raton, Fla.: CRC Press.
Anyone who wants to progress beyond the Davies equation, or whatever data may be found in *Stability Constants* at non-zero ionic strengths, will find the discussions of theory thorough, and the data on systems related to natural waters (an article by Whitfield) useful.

Riley, J. P. and Skirrow, G. *Chemical Oceanography* (4 volumes). New York: Academic Press.
See especially Chapter 9 of Volume 2 (1975), by Geoffrey Skirrow, on carbon dioxide and its relationship with seawater. The appendix to that chapter also contains some useful tables.

Sillén, Lars Gunnar and Martell, Arthur E. *Stability Constants*. Special Publications No. 17 (1964) and 25 (1971) of the Chemical Society, London.
Equilibrium constants for most ionic reactions that had been studied up to 1970 will be found in these comprehensive volumes with detailed bibliographic references.

Snoeyink, Vernon L. and Jenkins, David. 1980. *Water Chemistry*. New York: Wiley.
Those students who wanted to study from Stumm and Morgan but found it too difficult will be pleased with this book. It follows the same outline, but gives more details and examples, and omits the complicated aspects with good judgment.

Stumm, Werner and Morgan, James J. 1981. *Aquatic Chemistry*. 2d ed., New York: Wiley.
A comprehensive account of physical chemistry applied to natural and polluted waters. The chemist or engineer will find it full of unusual viewpoints. An entire chapter is devoted to the carbon dioxide system, and another to the regulation of natural water composition.

Index

Index of
Equation Numbers

*Equations on pp. 116–118 numbered (5.18)–(5.20) are different from equations with the same numbers on pp. 113–114 and should have been numbered (5.21)–(5.23), except that other equations already had been assigned those numbers.